# THE SEVEN SIMPLE STRATEGIES OF J



*Say What You Mean and M*

### Expect Greatness

Teach your kids that greatness is within their grasp. Expect it and it will come; reach for it and it will instill hope, dedication, and purpose.

### Say What You Mean and Mean What You Say

Choose your words carefully when you speak to your children, and be consistent in all things at all times.

### Listen Carefully

Be there in a physical sense as well as emotionally, spiritually, and every other way besides. Don't simply go through the motions. You can't ask your kids questions and not listen to their answers.

### Keep Your Word

Your kids need to trust you. If each and every one of us kept to our word in matters big and small, there'd be no such thing as neglect, disappointment, or false hope.

### Cheer

At all times, it's important for our children to hear us cheering them on, to get the message that we're here for them and pulling for them.

### Make Money Matter

We must pass on to our children a clear understanding of, familiarity with, and respect for money.

### Reach, Teach, and Preach

In one way or another, we must be accountable, and we must teach our children to be accountable, whether to some higher being or to some sense of civic-mindedness or to simply doing the right thing by the right people.

"Our children are watching the messages we are sending to them."
—Judge Glenda Hatchett

## *About the Author*

JUDGE GLENDA HATCHETT served eight years as judge of Fulton County Georgia Juvenile Court. She is Georgia's first African American chief presiding judge of a state court and was the department head of one of the largest juvenile court systems in the country. She developed partnerships with community organizations and businesses, including the Boys and Girls Clubs, the Urban League, and others, and helped found the Truancy Intervention Project.

After her first year on the bench, Judge Hatchett was selected as "Outstanding Jurist of the Year" for her pioneering leadership by the National Bar Association's local affiliate. She was also honored with the Roscoe Pound Award, the highest award for "Outstanding Work in Criminal Justice" from the National Council on Crime and Delinquency. The Spelman College Board of Trustees selected Judge Hatchett to receive the Outstanding Community Service Award, and she was honored with the NAACP's Thurgood Marshall Award. Nationally recognized as an authority on juvenile and social issues, Judge Hatchett frequently delivers speeches and lectures throughout the country. She is the mother of two sons, and splits her time between Georgia and New York.

# SAY WHAT YOU MEAN *and* MEAN WHAT YOU SAY!

*Saving Your Child from a Troubled World*

JUDGE GLENDA HATCHETT

Perennial Currents
*An Imprint of* HarperCollins*Publishers*

A hardcover edition of this book was published in 2003 by William Morrow, an imprint of HarperCollins Publishers.



HarperCollins books may be purchased for educational, business, or sales promotional use. For information please write: Special Markets Department, HarperCollins Publishers Inc., 10 East 53rd Street, New York, NY 10022.

First Perennial Currents edition published 2004.

*Designed by Renato Stanisic*

The Library of Congress has catalogued the hardcover edition as follows:

Hatchett, Glenda.
Say what you mean and mean what you say! : 7 simple strategies to help our children along the path to purpose and possibility / Glenda Hatchett ; with Daniel Paisner.
p. cm.
Includes bibliographical references.
ISBN 0-06-056308-7
1. Child rearing. 2. Parenting. 3. Parent and child. I. Paisner, Daniel. II. Title.

HQ769.H383 2003
649'.1—dc21 2003054122

ISBN 0-06-056309-5 (pbk.)

04 05 06 07 08 ❖/RRD 10 9 8 7 6 5 4 3 2 1

*In honor and memory of my father, Paul L. Hatchett, Sr.*

*In honor of my mother, Clemmie Barnes Hatchett*

*In celebration of my sons,*
*Charles Spurgeon and Christopher Lawrence*

*Your children are not your children.*
*They are the sons and daughters of Life's longing for itself.*
*They come through you, but not from you,*
*And though they are with you, yet they belong not to you.*
*You may give them your love but not your thoughts.*
*For they have their own thoughts.*
*You may house their bodies but not their souls,*
*For their souls dwell in the house of tomorrow, which you cannot visit, not even in your dreams.*

—KAHLIL GIBRAN,
*The Prophet*

# Contents

## Opening Remarks

# Promises to Keep

I was not a perfect child. I am not a perfect parent. And my children aren't perfect either. I've made a whole bunch of mistakes, on both ends of the parent-child relationship, but I've tried to learn from my mistakes—to the point where I'm at least a little less likely to make the same mistake twice.

Growing up, I learned a whole lot from my mother and father about the kind of parent I wanted to be, and I continue to learn a whole lot in my follow-your-heart, go-with-your-gut, on-the-job training as a mother.

For the most part, I've got the parenting thing and the professional thing and the personal-time thing down to a fine balancing act, even though sometimes all those balls in the air have to keep themselves from falling. I wouldn't change a piece of who I am or what I do or how I've chosen to go about doing it, except that I would pay more attention to taking care of me: finding more time in my hectic schedule for exercise, more time for long walks on the beach, more time for nothing much at all.

All of which begs the obvious question: Just who am I, and what is it that I do that has propelled each of us to these pages? Well, I'll get to all that, in some detail. For now, I'll offer the thumbnail version. I'm a single parent. I'm a lawyer. I'm a judge—a real-world, heart-weary juvenile court judge who has lately presided on a syndicated daytime television courtroom program that bears my name and seeks to make a difference in the ways we think about responsibility and community and kids in trouble. I'm an advocate for children, working women, and family court issues. At work I have been the repository for every manner of parenting misstep you can imagine, and a few more you couldn't possibly believe, while at home I've sought to take positive strides to ensure that my children and I don't wind up in somebody else's courtroom.

The book you now hold in your hands is all about my efforts to get it right. It's about all those positive strides I've worked to take as a parent and all those horrible missteps I've seen as a juvenile court judge. It's about what we parents can do to guide our children along a positive path, even as they are being pulled in so many conflicting directions. Also, and significantly, it's about the stories of my professional life. Gang violence. Drug abuse. Arson. Neglect. Assault. Murder. As a juvenile court judge presiding over one of the busiest jurisdictions in the country, I've seen it all, and a little bit more besides, and when you reduce each case to its component parts you begin to see a pattern. There's usually a well-meaning parent who can't quite think his or her way through a looming trouble. There's usually a confused child who doesn't have the support or the guidance to get past that looming trouble. There's usually an overburdened system, like a school or a family services program, that can't seem to find the resources to diffuse that looming trouble. And there are usually other looming troubles to combine with the first looming trouble to make matters worse.

This is how it goes. Sometimes. And sometimes, with the right mix of love and faith and tolerance and support and mutual respect and patience, situations can turn in other ways. In positive ways. In hopeful ways. And that's what this book is all about—helping each other to help our children walk the path of purpose and possibility. Drawing on our shared experiences so that our children can make better decisions and so that we can support them with better decisions of our own. Finding the resolve and the reason to give our children the earned benefit of the doubt, to the point where we can't help but come down on the side of hope.

My father used to speak about the crossroads in a young life, about turning the corner onto a street he called New Hope Road. As a kid, I wasn't entirely sure what he was talking about; as an adult, I am absolutely clear on it.

"You've got to turn the corner, Glenda," he'd say, and

I'd nod as if I knew what he meant, but it wasn't until I was a parent myself that I truly got it.

What he meant was that if you're really committed to changing your life, to walking a better road, all it takes is a change in direction and a commitment to that change in direction. Turn the corner, and keep walking, and after a while you look back over your shoulder and you can no longer see the old road. After a while, that new path—New Hope Road—becomes the only way.

I always wanted children. Indeed, I earnestly and repeatedly prayed for children, and God answered those prayers and blessed me beyond measure with two wonderful sons. I had fully expected to love my children, but I never understood how magnificent and magical the relationship with my children would be, and how it might grow even more magnificent and magical year after year after year. I have often said, and I make no apologies for it, that my priority in life is to give to the world two strong, loving, anchored, sensitive, purposeful, caring, focused, and successful young men—men who in their own way and in their own time will touch the world and leave it a better place for their being here. It's a lofty goal, I'll admit it, but I'm working on it. Actually, *we're* working on it, my two boys and I, and I'm guessing that if you've reached for a book like this, it's something you're working on too. In your own way. In your own time.

In the very last conversation I had with my father, two days before he died, he turned to me and said, "Glenda, you take good care of my grandsons, you hear?"

It wasn't at all unusual for my father to part on a loving, positive directive such as this—indeed, it had become his custom to send me off with a charge regarding his grandchildren—but it strikes me as somehow poetic that those were his absolute last words to me. And poetic too that my last words back to him were a promise in answer: "I will, Daddy," I said.

I have tried to keep that promise—and this book is a reflection of that. It's a distillation of my dovetailing perspectives as a juvenile court judge and a mother of two, out of which there emerges a hopeful path. Yes, absolutely, things can go horribly wrong raising kids in these uncertain times, even for the best-intentioned parents. But I'm here to tell you things can also go wonderfully right—and when they do, they usually do so for a reason.

And so I present seven simple strategies for parents hoping to raise smart, safe, successful children—each bolstered by bulletins from my juvenile court bench and reflections from my own kitchen table. Realize, these bulletins have been pulled directly from real-life cases, involving real kids in real trouble, and I've naturally taken some real precautions here; names have been changed, and in some instances composites have been used to further protect the identities of the children and families involved. If any of these strategies are appropriate to your own situations with your own children, use them and be blessed, for our children are all we have with which to build a new generation of men and women. Let's give it our best shot.

## Simple Strategy #1

# Expect Greatness

Allow me to repeat myself: expect greatness. Set the bar high. Encourage your children to reach beyond their wildest dreams.

There, that gets us past the easy heading, but I can't move on without overstating what may or may not be so obvious: it's not the greatness that matters, but the expectation of it, the reaching for it, the setting it out as a goal or ideal. Let's be honest, even the greats stumble from time to time, but greatness is not the point. It's the *prospect* of greatness, the preparation for it, the willingness to let it into our lives. Expect it and it will come; reach for it and it will instill hope and dedication and purpose and all kinds of great things; take it for granted and it will always slip away. It has to do with destiny, don't you think? If we teach our kids to move about as if greatness is within reach, then it will be so. If we teach them to hang their heads and despair about ever reaching their objectives, then they won't.

We parents must demand the utmost of our children, whether we're teaching them to tie their shoes, or read, or drive a car, or develop good study habits and a responsible moral code. And why stop at parents? Teachers, mentors, bosses, aunts and uncles . . . *judges*. We should always expect the very best from our charges if we hope to see the very best in return. It's basic. And just to flip it around, we must also expect the very best from ourselves—because after all, we set examples by our actions. You can't lead where you aren't willing to go.

No mother is certain about what she's doing when she first has children. In a world of uncertainty, this is one absolute. Me, I wasn't clear on a whole bunch of things, but I had some ideas, and one of the big ones was that I should give my kids something to shoot for, help them walk a path that might allow them to discover their dreams and realize them as well. In so doing, naturally, I set the bar high for myself, but I figured my children had a right to expect a kind of greatness from me if I was expecting it of them. It's only fair, don't you think?

I got this notion from my own parents, who nurtured and inspired my brothers and me to reach beyond our circumstances, but as hard as my mother and father worked to instill in me a sense of boundless opportunity and wonder, there were others working just as hard to tell me what I *couldn't* do, where I *didn't* belong, and when I *shouldn't* even bother trying. For every piece of positive reinforcement I took in at home, there were a dozen negatives out there in the rest of the world, and I came away thinking I'd have to level the playing field a bit when it came my turn to raise children. I didn't want them to have to deal with all these disapproving messages if they didn't have to, at least not at home, at least not on my watch.

It's funny the way a lot of my well-meaning friends and relations lined up to tell me how much trouble my boys would give me when they reached adolescence, based on how much trouble their own children had been giving them, especially boys. It's as if they needed to drag us down into whatever it was they were struggling through, either because misery loves company or because they couldn't bear the thought that they might have missed an opportunity to set things along a more positive path with their own children. "Enjoy them now," I'd keep hearing when my children were young, "because there's gonna be war in your house when they get older."

Each warning was more ominous than the last:

"Just wait till they become teenagers."

"There won't be peace in your house till they're both off to college."

"I don't envy you. Two boys. You're in for some real trouble."

I refused to claim this mind-set as my own. I flat out didn't want to hear it. Why? Well, if you go in anticipating a negative outcome, the positives won't know where to find you, so from the very beginning I turned a deaf ear when someone tried to counsel me on what almost everyone assumed would be problems in my relationships with my sons. I wouldn't listen. When there was no avoiding it, when someone needed to download her troubles and let out a little steam, I nodded politely until she was through. I'd half hear these terrible things and promptly set them aside, never once believing that any of these worst-case scenarios would have anything to do with the positive relationships I'd carve with my own adolescent and teenage sons when the time came. Those negatives weren't mine, I told myself; they have nothing to do with me or my children.

There's an old adage in the world of sports that suggests that great teams sometimes play down to the level of their lesser opponents. Conversely, there are lesser teams—less talented, less physically gifted, less dedicated—that play *up* against tougher competition. I've always believed that in cliché there is truth (sometimes, at least), and at this point I've seen enough high school football and basketball games to know that the adage applies, but I could never accept allowing my children the wiggle room to play *down* to expectations. What I mean by this is that I wouldn't give my kids an out, or an easy excuse. Better they should play *up* to what might at first blush appear a too lofty objective. Better they should aim high and push themselves above and beyond their easy targets. Every day, in my courtroom, I'd see parents who never expected a lick from their kids, and I'd come away thinking that these folks had chosen their own path and made it all but impossible for their children to do anything but drop the ball. What did these parents expect, tuning out on their kids, leaving them to fend for themselves, looking the other way when their kids were transparently in need of a push in a better direction? Of course these kids messed up, because they were held to such a low standard at home, because so little was expected of them, because their own parents put so little effort into setting things right in their own quiet corner of the world. It was more likely than not that these kids would screw up, and I was endlessly amazed by the number of parents who were actually surprised by this outcome.

I should mention here that a great many children achieve their own kinds of greatness without any help at all from their parents or guardians. They manage to thrive despite the negative influences on them at home and at school and wherever else they might care to look. Good for them. *Great* for them, actually. But these children are the exception, and whenever I hear the accomplishments of one of these high-achieving-despite-no-support-at-home kids, I catch myself marveling at how high they might have climbed with parents who demanded their best efforts, at all times. My hat goes off to these kids for what they've managed on their own, even as my heart goes out to them for what they might have missed—because after all, children should strive *because* of the adults in their lives, not *in spite* of them.

Here's a story from my own household to illustrate. My younger son, Chris, is a good basketball player. Actually, he's better than good, but it won't do for me to be bragging on him in these pages, so let me state it plain for the purposes of the story. Chris has got a strong, all-around game—strong on defense, strong on the boards, strong to the hole. About his only weakness, like Shaquille O'Neal's, is his free-throw shooting. It's his Achilles' heel. In this one area, Chris is not much better than hit or miss, to the point where opposing coaches know that he's the guy to foul when the game's on the line. To his credit, Chris recognizes this flaw in his game, and he's worked on some of the other weapons in his arsenal to compensate, but he knows that no matter how hard he works on his free throws he might never have that sure, soft touch from the line that some of his teammates seem to have been born with.

Okay, so that's the back-story. Front and center, there was Chris's high school team, time running out, meaningful game against a key rival, the other team up by a single point. Our guys had the ball and someone sent it down low to Chris, who was promptly fouled. There were just a couple of tics left on the clock. Chris went to the line for a one-and-one—that is, if he made the first free throw, he'd have a shot at a second. The crowd and the players on each bench were stone silent. Actually, it might have just appeared that way to me, with my son on the line, the game on the line, but it really seemed that everything else in that gymnasium had been put on pause. There was just Chris, stepping to the foul line, getting ready to shoot.

He sank the first shot, to tie the game, and from my

spot in the bleachers I could see him draw in a deep breath. His deep breath matched mine. He'd gone from everything-to-lose to everything-to-gain in the sure, soft flick of his wrists, and I could see the difference in his demeanor. Don't misunderstand, he was confident before taking that first shot. There's a swagger to him, come game time, but a mother knows how to look past the swagger. The thing about Chris is that he's always confident, always believes in himself—or never lets it show when he isn't, or doesn't. I guess maybe I'd been imagining how the moment might have been weighing on him and figured it must have been a big relief to make that first basket, because the worst thing would have been to miss the shot and have the rebound fall into the opponents' hands as time ran out and the game slipped away. This way, with the game now tied, at least Chris wouldn't be the scapegoat. Now all he had to do was grab his chance at being the hero. It wasn't exactly a no-pressure situation, but there's no question that there was less pressure on this second shot than on the first.

He took his time before taking his second shot, let the gym return to quiet, then sank the next basket as if it was the most natural thing in the world.

The crowd went wild. Chris's teammates went wild. His poor, crazy cheerleader of a mother went wild. (You want to talk about pressure, try being the mother of the player on the foul line in such a tense situation!)

Those two foul shots put our guys ahead by one point, and as time ran out and Chris was swallowed up by the pandemonium of the moment, he quickly broke from the backslaps and congratulations of his coaches and teammates and sought me out across the gym. He had the ball in his hand, and he ran over to where I was sitting and handed it to me. (Goodness, he nearly leaped across the court and over to my seat, that's how excited he was!) Of course, giving the game ball to his mother was mostly for dramatic effect, because there's no high school in the country about to let a kid give away school property to someone in the stands, but I wasn't going to quibble. It was, I thought, a grand gesture, and I received it with a full heart, my eyes welling with tears of pride and joy and (okay, I'll admit it) . . . relief. I'd give the ball back later, when the coach came looking for it; for now, it was mine.

Chris collected me in a great big sweaty hug. "Mom," he shouted, above the din, "you and I were the only people in this gym who thought I could hit those shots! We showed 'em, huh?"

I hugged him back and I thought to myself, Yes, Chris, we did. *You* did. *You* showed them. You stood there knowing that your mother believed in you wholeheartedly, and you took that belief and turned it on yourself. In the end, that's the basic message: expect greatness or fall short.

How remarkable was that? Really, I was so drop-dead excited, so over-the-top proud, I was about to burst. But it wasn't just the action that had me reeling or the result of the game; it was the fact that Chris had seen so clearly what I'd been reaching for all along. Plus, he'd managed to give it voice. To have your child seek you out at such a moment, to publicly thank you for your unwavering belief in him, to claim that belief for himself . . . man, that was something! Yes, it was just a game, but it got me to thinking that it was more than just a game at just that moment. It was a validation of what I wanted for my children—to be able to stand on their own two feet and meet whatever challenge lay in wait with confidence and hope. And to do so knowing their mother was on the sidelines cheering them on with more of the same!

Clearly, the percentages don't *always* favor such picture-perfect outcomes, and Chris knows that I would have hugged him just as hard if he had blown it and his team had lost the game. (Harder, probably.) I would have told him he'd given it his best shot, and that I was proud of him, and that there would be other opportunities. And I would have meant every last word. There have been a ton of times I've had to hug away a disappointment when things didn't go quite this right, but that's all part of the same package. Realize, in these cases it's my son's disappointment, not mine. The only time I'm disappointed is when the effort isn't there. But because we've set such high standards for each other in our family, and because Chris and his brother, Charles, have always been encouraged to shoot at a different level, they come out at a different place. Does that make any sense? Chris didn't sink those free throws *because* he'd always been told that his mother expects great things from him, but the cumulative effect of all those great expectations allowed him to dream possible dreams, to stay within himself and reach for every imaginable possibility. And he didn't seek me out across that gymnasium to gain my approval. He's got that with him all the time. No, I like to think he sought me out to validate the belief I have in him, the belief he now has in himself.

It wasn't a fluke, Chris going two-for-two with the game on the line like that. It wasn't dumb luck. It was the by-product of a lifetime of encouragement and nurturing and cheerleading and discipline and love—pure, unconditional love. Actually, to put a fine point on things, it's unconditional love with just a few strings attached. I don't believe it's enough to love a child and leave it at that or for that child to receive love and give nothing but his own love in return. No. It starts there, in our house, but it doesn't end there. In my book it says to expect the best from those you love and to offer your best in return.

Now to the bigger picture . . .

In my courtroom, almost every day, there was evidence of children being raised with no expectations at all. When Terry Walsh of Alston and Bird and I first launched the Truancy Intervention Project, a multilayered initiative to combat truancy in our community's schools that integrates computers and teleconferencing software into our follow-up effort, I was exposed to large groups of underachieving Atlanta teenagers whose only crime seemed to be that they had failed to exceed the low expectations set out for them by their parents, or their teachers, or their community at large. In one such group, there were thirteen girls, middle school to high school age, and I made it our shared mission to turn their lives around. Then (and still) there was a direct correlation between the truancy statistics and the numbers tracking teenage pregnancies; girls who repeatedly ditched school were far more likely to become pregnant before leaving high school than their more conscientious peers. And so I set it out as a challenge. I told these girls that I would not tolerate a single pregnancy among our group for the balance of our time together, which would take us through the entire school year. Plain and simple. We've all read about zero-tolerance policies in our schools and in our courts, but on this one point I was subzero. These students came from families with a history of teenage pregnancy; their older sisters were having babies at fourteen, fifteen years old; in some cases, their own mothers had had them when they were still children. It was time to break the cycle.

We'd meet regularly throughout the school year, and there were caseworkers and social workers tracking their progress when I wasn't around, and these girls knew I meant business. Like I said, my subzero tolerance left no room for doubt; my expectations were clear. And do you know what? Not a single one of these girls was pregnant at the end of the school year.

Can I draw any great conclusions from that one small group? Absolutely not—at least not any statistically significant conclusions. But I can draw a line from the expectation to the realization. I can look back at these girls, and the hard line I took, and see a connection. Plus, I didn't merely lay down the law and expect these girls to fall in behind it. I gave them something to work with. I asked them, one by one, "What is the dream you have for your life?" And I listened to their answers.

One girl muttered shyly that she wanted to style hair. Again, it was a place to start, but I let her know her answer wasn't complete.

"Tell me you want to be a cosmetologist," I urged. "Tell me what it's gonna take for you to get your license. Tell me what it's gonna take to start your own business."

By the end of the conversation, I had this girl thinking she could finish high school, go to vocational school, get her cosmetology license, take some business courses, and find what it would take to open her own shop. We walked each other through the entire process. I told her I wanted to be driving down Peachtree Street one day and see her salons, and this girl was astonished. Really she was aghast, because I was expecting so much of her, more than she expected of herself.

We went on from there. "This guy you're with," I said, "he's pressuring you to have sex, but he's not gonna be there for you if you get pregnant. Think how hard it'll be to open your own salon if you've got a kid at fifteen."

In every case, I tried to get these girls to be willing to align themselves with choices that would help them realize their dreams.

Another girl couldn't say what her dreams were but allowed that she loved animals, so we turned the talk around to veterinary school. Another expressed the special feeling she got helping her younger sister learn how to read, so we explored the idea of becoming a teacher.

There was one eye-opening exchange after another, and in each case something wonderful happened. The world opened up, to the point where these girls were no longer looking at dead ends but at crossroads. None of them had ever thought about college, because no one in their families had ever gone to college. No one had ever told them such a thing was possible. But that was precisely the point. Anything is possible, I told these girls. If you set your mind to it, if you come up with a workable plan, you can do anything. I don't care what your family background is. If you find the right motivation, everything else can follow.

Expect greatness. (There it is again.)

In juvenile court, I got in the quick habit of telling kids who came before me that there was a dream out there with their name on it. It became one of my signature messages. And then I'd tell them the deal. "Nobody can claim that dream for you except you," I'd say. "It's out there, waiting on you. And nobody can get in your way. It's all on you. Your mother can't get in your way. Your father can't get in your way. Your best friend can't get in your way. And it can't be Michael Jordan's dream. That is uniquely his dream, so don't stand there and tell me you want to be Michael Jordan, because you will never be Michael Jordan. That job's taken. He's a unique, distinct individual—and so are you."

There is no cap on dreaming, but the only dream worth having is your own, right? Encourage your child to dream his own dreams—not yours, not his best friend's, but *his*, and his alone.

Case after case, day after day, I put the expectations back on the child. It wasn't for me to set the bar for them; the bar was set based on their dreams, on what was attainable, because as important as it is to dream, it's even more important to dream with a plan. We looked at the best-case scenarios for their lives and then looked higher. I encouraged each child to lift his or her vision. If their sightline was at this or that mid-level expectation, I urged them to raise it. Over time, the message became a mantra in my courtroom: "Lift the vision." Whatever the goal, it could be greater; wherever the bar, it could be higher; whatever the commitment, it could be stronger.

Sometimes a kid wouldn't have the first idea what he or she wanted out of life, and in those cases I sent them home with an assignment. If I set a sixty-day review, or a ninety-day review, I'd ask the child to spend some time thinking on it, to start the process of dreaming and planning and thinking in ideal terms. I'd order them to keep a diary to present to me upon review, on the theory that it is indeed important to set these "life plans" down on paper, to connect the dots between the stuff of our dreams and the stuff of our realities. It's one thing to have a goal and quite another to lay out a map to get you closer to it.

Each week, I expected these kids to write down one thing they did during the previous week to make their lives better, to bring them closer to their dreams. It got to the point where I could tell which kids took the task seriously and which scribbled hastily to fill the empty pages the night before their review, but even in these cases the effort was better than nothing. You'd be amazed at the change in kids' attitudes when they think someone believes in them; and when that someone is a parent and that belief is reinforced every day, in every way, there's no telling what a kid can achieve. It's not just the belief that's key—it's the constant reminder that someone is watching them, that someone cares, that someone is there to hold them accountable.

So yes, absolutely, expect greatness. Set the bar high. Push your kids to opportunities you never had yourself.

I heard a case early on in my career that shows what can happen when even the best-intentioned parent doesn't take the time to figure out a child's wants or worries. An adolescent boy was brought in on shoplifting charges. I rarely heard misdemeanor cases, but on this one occasion there was a mistake in my calendar, and rather than reset the date and put these good people through these difficult motions a second time, we went ahead. Here was this little boy who'd taken a bunch of tapes from the local video store while waiting for his adoptive mother to complete her shopping elsewhere in the mall. He stuffed the tapes under his jacket and made for the door. The scenario didn't make a whole lot of sense. The mother had sent the kid off to do his browsing with the promise of collecting him there and paying for his rentals or purchases. But as we spoke in my courtroom on the day of the hearing, an explanation presented itself. This was a child who'd been abandoned by his biological mother and adopted by an empty-nest couple. The adoptive mother had worked in a homeless shelter, befriended the boy, and convinced her husband that they should make a home for him. The boy was about nine or ten years old at the time of the adoption, about twelve or thirteen at the time of the video store incident, and by all appearances things had been going well. The boy was loved. The parents had provided a comfortable and supportive home. The boy was doing well in school and interacting appropriately with his peers. Underneath the veneer, though, the child was struggling. He missed his biological mother. He'd never met his biological father and wondered if there was anything he might have done to drive the man away. He worried that his adoptive parents would soon abandon him as well.

It really was a heart-wrenching case—but they're *all* heart-wrenching cases, aren't they? Here, the boy couldn't really articulate what was troubling him, and he chose instead to act out in this public way. It was almost as if he went into that store determined to put his adoptive parents to the test, to see if they'd still be there for him if things got rough. He was a smart kid. He knew there were sensors on those videos. He knew he'd be caught. He did this for a reason. Did he think these things through on any kind of conscious level? Probably not. But somewhere, deep down, he knew what he was doing. And somewhere, even deeper down, in a place he didn't entirely trust, he knew his adoptive parents would be standing right behind him, no matter what.

There was another piece to the puzzle, which became clear when I talked to the young man in my chambers, without his adoptive parents present. For all the love and attention these good people had showered on this child in the four or five years he had been in their home, they never really took the time to push him, to steer him down a path where anything extraordinary was required of him. They hadn't expected greatness from him. Not even close. They hadn't expected anything at all. They were kind, well-intentioned, loving people, don't get me wrong; they were crying in my courtroom when the charges were read. But they'd been so concerned with providing for their adopted son, in making up for the harsh, neglectful treatment he endured in the early part of his life, that they hadn't taken the time to motivate him, to give him something to shoot for, to set certain standards. It turned out that the kid had a passion for tennis—and he was supposedly really good at it too. But he didn't play in any kind of organized program, and what surfaced in our session was that this was the kind of kid who could thrive with the structure and discipline that concentration on an individual sport might provide. He desperately needed that something-to-shoot-for.

Another judge might have heard this case and sentenced this young man to a juvenile detention facility, but I couldn't see the point in locking him up with a bunch of kids who could turn him on to drugs and gang activity. That punishment didn't fit the behavior. This was an inwardly troubled kid who in every other outward respect managed to move about successfully in his new environment. He made good grades. He stayed out of trouble. He ran with a (mostly) good crowd. I chose to look on this one misstep as just that—a misstep, and a cry for help. It was also a warning, sent consciously or subconsciously to alert his adoptive parents that something was wrong with this picture. Something was missing.

So what did we do about it? Well, I helped the family arrange for some counseling, which I felt was key. There were some desperately important issues that this young man needed to work through concerning his biological parents, and he needed the support and encouragement of his adoptive parents to do so successfully. It's not uncommon for adolescent children—boys, especially—to get really stuck on the notion of who they are and where they come from, and sometimes there's no way to get unstuck without professional help. So in this one case the counseling piece was all-important. I also sentenced the child to a modest amount of time in community service—at a local homeless shelter, which I felt would serve as a not-so-gentle reminder of how things sometimes go for kids from similar backgrounds who haven't been as fortunate as this young man in finding a loving adoptive home.

The final component came with a call I made to a tennis pro I knew, because without a separate motivational push I didn't think the counseling or community service would be enough to turn this situation around. This young man very much needed a goal in his life—a clear, attainable goal. For his first eight or nine years that goal had been merely to survive, to string one day to another and somehow manage; at some point, after he was left by his biological mother, that goal shifted slightly, to where his focus was on finding an adoptive home. I don't mean to diminish what this child went through, because no child should have to face what he faced, but now that he had landed on his feet, on solid, loving ground, he needed another shift in focus. And his parents needed to help direct that shift, to help him find a place to put his energy and creativity on the line. At that moment, that place might be on a tennis court, so we arranged for a round of lessons. At some other point in his life, it might take him someplace else. But at every point, for every child, there should be an outlet for greatness, an opportunity to pursue something they're really good at or even something they think they might be good at. A place to push themselves to be the best that they can be, with parents at their side expecting nothing less.

A place to claim the dream with their name on it.

Another case. A hard case. A young woman, fifteen years old, had been in and out of my courtroom back before I could even call it mine. Goodness, this child had done everything. Drugs. Prostitution. Shoplifting. By the time I reached the bench, she was knee-deep in the system, with a case file thicker than an encyclopedia, and when she surfaced on my calendar a second time, I'd had enough. Her name was Lisa, and it was apparent from her file that she was bright, and after spending just a few minutes with her it was clear that her potential had not been unleashed.

"I am absolutely sick of you being here," I admonished her at the top of her hearing.

Later Lisa told her caseworker that I threatened to

come down off my bench and wring her neck if she came back into my courtroom on a new charge. I don't remember saying that, but I guess I did. This poor child was one of hundreds of girls, in hundreds of versions of the same bad way. If I didn't say it I should have, because this girl needed a talking to—and for now, anyway, it fell to me to do the talking.

"You should be getting ready to go to college," I finally said. "You're out there acting the fool, getting into trouble, and you should be doing something with your life. You're better than this."

"Yes, ma'am," Lisa said, in the kind of sheepish voice these girls sometimes used when they were caught. (Or when they were confronted by a menacing judge who was threatening to wring their necks!)

I was angry at this teenage girl—not just for clogging up our courts with such repeat nonsense but far more importantly for throwing away whatever opportunities she might have had. Really, I can't abide such wastefulness. This child had no concept of what it meant to strive for greatness and was only too happy to keep doing what she was doing, and her disinterest set me off. But I couldn't let that anger get in the way of the task I'd set out for myself—to give these kids a goal, something to shoot for, some standards to meet. If there were no parents around to expect the best out of them, I'd cast myself as a surrogate.

"What is your dream?" I said for the zillionth time, trying not to let exasperation show. "If you could do anything in the world, what would you do?"

Lisa's answer floored me and just about everyone else in my courtroom that morning. "I want to be a doctor," she said.

All right! I thought to myself. Now we've got something.

Realize, no one in that courtroom was prepared to take this girl seriously. No doubt some folks thought it was cruel of me to put her through these paces, to set her up for another disappointment. This was a child who was barely going to school, who'd been brought in on drug and prostitution charges more times than any of us cared to remember, and yet there she stood, straight-shouldered, looking me directly in the eye, declaring that she wanted to be a doctor. How could we *not* take her seriously? Really, it took great guts to come out and make such a statement, and I would not hear it lightly.

"Okay," I said, "so what's it gonna take to shake some sense into you and get you on track to become a doctor?"

What it took, I learned later from the caseworker, might have been that tossed-off threat I made from the bench, which got her to realize that she had gone perilously wrong, but whatever it was, she now stood before me, contrite and vulnerable and willing to work hard to set things right. After all was said and done, she had me convinced she could do anything she set her mind to, and I told her as much. "I absolutely believe you can do this," I said, and we established a blueprint we'd follow together, to help her reach her goal. School. Probation. Caseworker visits. A close review schedule. Counseling. The whole deal. Locking her up at this point would have had no impact. There was no shock value to it any longer. She'd have probably known half the girls in the detention center anyway. They'd greet her with a "Hey, girl, what's happening?" and she'd fall back into the same bad habits. No, the thing to do was to move her up and forward, to set the bar high and help her to reach it.

I wound up placing her in an innovative outreach program called ALPP—Alternative Life Paths Program—which offered a transitional support network for rootless kids, provided the follow-through they needed in school, and offered independent, group home living situations for high school kids. Under the umbrella of ALPP and the watchful eye of its director, Camilla Moore, I knew Lisa would have a shot at getting back on her feet. There'd be some structure to her life and some expectations. She'd have a curfew and responsibilities. There'd be a real sense of discipline and purpose. For so many of Atlanta's troubled kids, the ALPP initiative offers a bridge to greatness, and I expected it would do the same for this troubled child.

For good measure, I reminded Lisa of all the strikes against her. "This is your last shot," I said. "Let's be clear. Either you turn it around, or I'm shipping you off to the state."

She'd been in the system enough to know that committing her to a state facility was the correctional equivalent of me stepping down from the bench to wring her neck. She didn't want to go there, and I didn't want to send her there, which put us on a kind of common ground, and

I sent her off with another of my favorite rallying cries: "I'm pulling for you," I said. And I was. I truly was.

Fast-forward a bit. As I write this, this remarkable young woman is finishing her undergraduate studies at one of the better schools in the state. Lisa hasn't yet been accepted to medical school, but she will be. I have no doubt. It hasn't been the easiest road, but she's kept to it. She made a couple of stops and starts at other schools before settling in at her current university. She lost her mother to alcoholism and alcohol-related liver disease a couple of years back. Her father stepped back into her life and brought with him another mess of complications. But through it all, she has persevered. She made an excellent personal connection with her caseworker. She's made excellent grades, earning dean's list honors every semester. And she's managed to hold down a part-time job, working at Kinko's at night. She even worked for me for a time, in an internship position. In virtually every respect, Lisa has turned her life around, and if you ask her, she'll tell you that the impulse flowed from that challenge in my courtroom. That's how she saw it, like a dare, like this know-it-all lady in a black robe was giving her the business, drawing a line in the sand, telling her it was time to step up or be gone. And—God bless her!—she stepped up. Big time. All because she finally had a plan for her life. All because someone believed in her. More to the point, someone expected something from her, which allowed her in turn to expect something from herself; someone expected greatness and challenged her to reach for it.

We've kept close. With graduation looming, Lisa stopped by to see if I'd attend. She didn't have to ask, but I was thrilled that she did. I was going anyway. And I told her so, in the shorthand I've adopted for all my "kids" who've managed to turn their lives around after an appearance in my courtroom.

"New hat, new dress, and a new pair of shoes," I said, meaning I'd go out and buy myself one of each for the occasion.

Goodness, if I had a new hat, new dress, and new pair of shoes for every time I've made this pledge, I'd be stylin' with "Sunday best" outfits. When it comes down to it, kids will come through, more times than not. Oh, yes, they will. Believe in them, and they can believe in themselves. I don't care if it's something as fleeting as finding the calm and the confidence to sink a couple of free throws in a tight spot, or if it's something as everlasting as picking yourself up off the streets and going to medical school. It all flows from what you require of them. Ask nothing, and you can be pretty sure you'll get nothing back in return. Ask everything, and anything is possible.

## *Side of Hope*

My mother, Clemmie Barnes Hatchett, is an amazing woman: bright, energetic, confident, adventurous, generous, creative, resourceful, independent, feisty, and passionate about life.

My father, Paul Lawrence Hatchett, was a wonderful man: smart, proud, kind, honest, wise, honorable, patient, dependable, fair, and determined to live life with a hopeful view.

Both my parents grew up in small towns in the Deep South—my mother in Florence, South Carolina, and my dad in La Grange, Georgia, hard by the Alabama border in the central part of the state. Both were capable of becoming anything they wanted to be. My mother dreamed of being a pediatrician; my dad dreamed the dreams of an entrepreneur, although he rarely gave them voice. Yet the reality is that they were both born poor, and they were both born into the grips of a segregated South, only two generations removed from the Emancipation Proclamation. They grew up in the throes of abject racism and sexism and every other kind of -ism that worked against you, where even educated young men and women were all too often denied opportunities commensurate with their abilities. Had my parents had a fraction of the opportunities I have had, my mother would have gone on to be surgeon general, and I always thought my dad would have been a phenomenal lawyer and an even more outstanding jurist. I am convinced that he would have served this nation with great distinction on the Supreme Court. And so the world lost out on their respective talents because of the insidious nature of discrimination. Those who were discriminated against were hurt, but the world also lost out in ways that we will never be able to measure or fully comprehend.

Instead of a pediatrician, my mother became a teacher, truly dedicated to her students, and later an assistant high school principal. And even though my father had done some graduate work toward his MBA, he was made to spend too many years of his working life toiling for the United States Post Office, reporting to a white supervisor who did not even have a high school diploma. Ultimately, after federal legislation forced open some doors that had been closed to him as a young man, my father was promoted to positions in federal agencies, eventually retiring from the Department of Labor.

I am old enough to know some of the injustices of that same segregated South. As a child, I knew what it was to be counted out before I could be counted in. I knew what it was to be called "nigger" by the white kids in town. I knew what it was to shoulder the taunts and inequities of that time and place, and my parents took turns teaching me to rise above them. My mother taught me that no one could put a name on me. "Glenda," she'd say, "nobody can tell you who you are but you. If it doesn't apply, then you can't respond."

And my father? Well . . . he taught me to carve my own path, and he did so with a simple, powerful message. The message had its root in a particular piece of discrimination that another parent might have ignored. You must realize, in the segregated schools of the Deep South, we got the hand-me-downs. The hand-me-down books and lesson plans, the beat-up and carved-up desks, the wobbly chairs, the rusted, dilapidated playground equipment. The entire operation was secondhand. In truth, I encountered many talented and inspiring teachers over the years, but in first grade all I knew was that I was bored. My mother had taught me to read before I started school, so I was reading ahead of the other kids, and doing simple arithmetic ahead of the other kids, and basically twiddling my thumbs while the clock on the classroom wall crawled to the closing bell so that I could race home and play with my friends and later get down to the real business of learning. In school, without any real resources, I was bored out of my mind.

Finally, midway through first grade, we got a set of readers handed down to us from one of the all-white schools across town. These books were torn and worn in every which way, but I remember being excited about them. We were all terribly, terribly excited. The teacher told us we would take turns reading aloud from our very own copies, and she handed out the books and we started in, going up and down each row, until every child had a turn. When my turn came, I was about ready to burst. At last, I thought, I'd get a chance to read out loud, to show everyone what I could do. But the page I was meant to read was torn in the copy of the book I'd been given. In fact, the entire page was almost completely ripped out. Some of the other pages had been taped back together; only this one was all but missing; so the teacher skipped over me and moved on to the next child.

I'd missed my turn, and it wasn't coming back.

I was devastated, and the day has been burned

into my memory to the point where it's never far from my mind. I can still picture what I was wearing—a brown dress, topped by a little white sweater with an embroidered rooster—and underneath the outfit beat the broken heart of a deflated little girl.

I went up to the teacher after class to talk about it, and she gently explained that all the books were torn, that it wasn't just my copy that was in such a sorry state, that each child was in a different seat on the same boat. I didn't much care about the other kids in the class, and I said as much. I said, "I don't care about everyone else's book. I want a new one. I want a book that's not messed up."

"Glenda," she said, "we don't get new books. You know that."

No, actually, I didn't. All I knew was that I didn't get a chance to read out loud to the other kids. All I knew was that my book was torn, with whole pages missing. All I knew was that it wasn't fair.

I raced home to tell my father. Surely, I thought, this was something he'd take care of. When you're six years old, your father can fix anything, right? Anyway, that's what I thought, and I set the situation out for him, trying to fight back tears as I told the tale. And do you know what? He didn't seem too terribly upset—at least not in the ways that mattered most to me, at just that moment. He wasn't concerned that I couldn't read about Dick and Jane and Spot, but at the same time he was troubled about something. What he cared about, really, was how I was processing my disappointment, and what I might learn from it, and to this end he sent me off with a paper and some crayons and told me to write my own story. I had a little red table in my room, with two matching little red chairs, and he sat me down there and told me I didn't need the discarded books from the white schools in town. The only things I needed were my paper and my crayons and my imagination; with these, I could write my own story.

Sad to say, my father's response didn't cut it—not at the time, anyway, not to a six-year-old girl. It wasn't until I was older, looking back, that I took in my father's message. He didn't care about Dick and Jane and Spot and how fast they could run. No. What was important to him, what he wanted me to consider, was how high I could fly. That was the real lesson and the real blessing of this exchange with my father. He couldn't fix our busted-up playground. He couldn't fix these taped-together books. He couldn't fix our racist society. But he could fix me. He could toughen me up and shore me up and strengthen me with the faith and courage and resolve to rise above any injustice. He could teach me to write my own story—and, indeed he did, in more ways than he might have imagined. The real lesson of that moment was that when you get to places in life's book that have been torn by injustice you can't stop, you must rewrite those broken passages and create your own story, in your own way, in your own time.

In hard places, in hard times, these were the messages that mattered. These were the lessons learned.

## Simple Strategy #2

# Say What You Mean and Mean What You Say

A grandfather gathers his grandchildren and offers them a challenge. He can't find his gun, he says. He only just had it, only just the other day, and now he can't place it, and he's thinking he can put all these kids to work to help him look for it. He makes it like a contest, offers a reward to the grandchild who turns it up, drums up enthusiasm for the search in what ways he can.

The children jump at the game, because that's how the old man sets it out, like a game, an adventure. They'd been bored, rattling around their grandfather's house on a rainy afternoon with nothing much to do, and here the old man presents them with this scavenger hunt, with the promise of a prize, even. Plus, it's not just a kid's game he's laid out for them; it's not like they're looking for a toy or a set of car keys; they're looking for a gun—a real-live, real-deal gun. They're off on important business, helping out their grandfather.

The cousins split into groups of two and three to begin their search. There are a half-dozen or so kids, ranging in age from eight or nine to thirteen or fourteen. The grandfather can never keep track, but he's pleased with himself for coming up with an activity to keep them all occupied—and in the bargain, maybe one of them will turn up his gun after all.

And that's just how it goes. The nine-year-old and the eleven-year-old come upon the weapon after just a few minutes. The cousins are like brothers—close in age, close in appearance, close in temperament. They do everything together. The older one picks up the gun and checks it out. He's handled a gun before. He thinks he knows what he's doing. He takes out the clip and pretends to shoot his younger cousin. It's all a part of the same game, his grandfather's game, but it's also a game he and his cousin have played out a thousand times before in their imaginations, with toy guns, or just pretending.

He's got no idea there's a bullet in the chamber. . . .

The most chilling part of this scenario is that it actually happened, and the eleven-year-old turned up in my courtroom charged in the shooting death of his nine-year-old cousin. Can you imagine? It was one of the most heartbreaking cases I ever heard, because it was such a senseless, ridiculous, tragically horrifying set of circumstances—

entirely preventable, if you ask me, if the old man had shown even the smallest amount of good judgment or the slightest ability to think ahead. And the fallout would tear this family apart. Really, there'd be no repairing what this grandfather had wrought with his stupid game. Step back and look at it: two adult sisters, one with a child dead at the hands of the child of the other. An eleven-year-old boy who for the rest of his life would have to live with this terrible truth—that he had shot and killed his young cousin, a boy who had been like a brother to him. And a nine-year-old boy, blown away by the horseplay and foolishness that, when you break it down, could not have shaken out in any other way.

Running through it all, there was this grandfather, who was nowhere to be seen in my courtroom. I might have known. His hands were all over this mess, and he'd turned tail. His daughters were there, doing what they could to hold each other up and move forward. His grandson was there, innocently facing the charges against him. There were other family members present, and friends all around, making the old man even more conspicuous by his absence. I still resent the fact that the district attorney didn't go after this guy on criminal charges. After all, it was his no-account foolishness that had set this whole heartbreak in motion. Sending all those children off to look for his loaded gun? What the hell was he thinking? And my anger and resentment weren't all at the district attorney. No, some was at myself. I should have hauled the grandfather into my court—at the very least to make sure he understood his culpability. But I didn't. I should have, but I didn't, and I regret that I didn't.

What I was able to do was put the eleven-year-old boy on extended, long-term probation—not because the child needed punishing, or to be placed on any kind of strict monitoring, but because he needed counseling, and the family didn't have the resources to get the child the kind of treatment I felt was essential. This way I could arrange it so that the boy and his mother and his aunt and the rest of those cousins could get the help they needed, on the state's dime, talk to the people they needed to talk to, and hopefully get past this traumatic ordeal. In its own way, the system provided for instances such as this, and I worked to make it happen; underneath, I knew these good people could never really get past these dreadful moments, but I felt I had to help them try to cope and move forward.

I set this story out as one of the most alarming red flags I've encountered to remind us how important it is to choose our words carefully when we speak to our children, to see situations clearly, from all sides, and to be abundantly and absolutely consistent, in all things, at all times. I'm still incensed at this grandfather after all these years but not because I ever thought he intended for this tragedy to happen. Let's give him the benefit of that doubt. It's not even because he was too stupid to recognize that this was how things *might* have gone. Let's assume he didn't have the tools to go out that far in his thinking. But there's no way I can excuse or even understand this man's directive, even if all those grandkids had spent the entire day looking for that gun and had never found it. I mean, the Children's Defense Fund has had some pretty harsh words to say to parents for not taking extra measures to conceal their licensed firearms and keep them out of the reach of children, and I don't imagine those folks would take a positive view of this man's sending his grandchildren off in search of his weapon—especially when it came to light that at least one of them knew enough to take out the clip before firing, a ritual the boy most likely learned at his grandfather's knee.

Clarity and consistency are all-important communication tools, and this is especially so when we talk to our kids. Now my hope here is that you look past the sketchy details of the shattered family above and see something of your own situation. No, I don't expect that you're out there putting your children directly into harm's way, the way this grandfather did, but there have undoubtedly been times when you've sent a confusing or conflicting message, when something you've said or done left the door open for a negative situation to come calling.

Let me pull back a bit and consider an illustration from my own household. Granted, it's a whole lot less consequential, but there are consequences to every exchange we have with our children, even the ones that don't end in tragedy.

Charles was four or five years old, and we were sorting through his old clothes, putting some of the stuff he wasn't wearing anymore in a box for Goodwill. It was, I thought, an important message for him to take in, even at that young age, that what we have in plenty we must also share, and as we considered each article of clothing, we talked about the little boy who might enjoy it now that Charles had grown out of it. It really was a sweet moment, and I remember thinking to myself how important it was to include Charles in the process. As messages went, I thought, there were none bigger—but then Charles went and flipped the whole scene on me, and in his own way he managed to reveal his mother as something less than clear (and a little two-faced, besides).

We came across a pair of raggedy old jeans that he had worn through at the knees, and as I instinctively moved to place the jeans in the box, Charles pulled me back.

"Mommy," he said, "we can't give these pants to a poor child." He pointed to the holes at the knees and finished his thought: "Just because someone is poor doesn't mean they have to be embarrassed."

Now it was I who was embarrassed, because of course he was right. I'd been so caught up in the mechanics of what I'd set out to do—sorting the items by season and size—that I failed to look at the scene from Charles's perspective. Of course, what he was seeing confused him. Of course, the mixed message of boxing his old clothes for charity didn't mesh with the sight of a worn-out pair of pants. Of course, Charles's observation made perfect sense and was far more consistent with the spirit of our giving than my piling the clothes into boxes.

The footnote to this moment was that Charles and I went to the fabric store, bought some patches, and together sewed them in place, and this time when we folded the patched pants and placed them neatly in the box, Charles was well satisfied that we were indeed giving some little boy something he could feel proud to wear.

So let's be clear: Sometimes it takes a child to bring a little clarity to a parent's perspective. And sometimes it takes a parent to pause for a bit, catch her breath, consider her child's view, and start over.

My own parents were endlessly clear and consistent in their dealings with us kids, and I have to think that their care and attention in this area informed the person I became. At least, they kept me striving. My own kids are growing up in a much more permissive, entitled environment, with a whole lot more going on in terms of outside influences and external impulses, so I've necessarily had to adapt. I honestly don't think a rigid approach would work in our home, in our community, at the turn of this particular century. Times have changed. Customs and standards have changed. *We* have changed. But what hasn't changed is the need for parents to be firm and focused, to hold out a set of reasonable expectations for their children to meet, and to hold them to it. Curfew is a good example. If you set curfew at eight o'clock and your child comes moseying in at eight-fifteen and you don't say anything, you've set a dangerous precedent. If the next weekend he wanders in at eight-thirty and you still don't say anything, you're headed for a clash, because at some point, when your child comes in at ten o'clock, or eleven, or midnight, and you finally snap, he'll be able to point back to your lax attitude and previous nonenforcement as a source of confusion. The responsibility will be yours as much as his.

I was never confused by my father's ground rules, and he was never lax about enforcing them. If he gave me a curfew, I was meant to keep it, down to the minute—no questions asked, no excuses heard. With my own kids, there's been some retooling. Here's an example: Charles came to me when he was old enough to drive and proposed that he have no curfew at all. I looked at him as if he'd lost his mind, and I said as much, but he pressed his point.

"Mom, it's not that crazy," he insisted, "just hear me out."

So I heard him out, and I hated to admit it but he made sense. He'd always shown good judgment, he was extremely responsible, I trusted him implicitly, and he ran with a good group of kids who for the most part displayed the same characteristics. Then he made some reasonable points: He reminded me that he'd be going off to college in two years (as if I needed reminding!), where he'd be on his own without a curfew, and he suggested we'd all do well to get him used to the concept now, while he was still on my watch. He also pointed out that a lot of times kids with curfews get themselves into trouble when they lose track of time and find themselves running late and maybe driving home a little too fast in order to beat the clock. Indeed, I'd heard a few such cases in my courtroom, so I knew that situations like these did occur, although I must admit those cases were few and far between. Of course, my response was invariably that they should have planned ahead to avoid having to speed home. Nevertheless, I listened respectfully to Charles's appeal and promised him I'd take it under advisement, which I did.

After careful consideration, I surprised myself by seeing things his way—at least on an interim basis. My father would have cringed if he'd lived long enough to see Charles with his curfew lifted in the summer after tenth grade, but I felt I owed it to Charles to give his proposal a dry run. He'd earned that much at least. And do you know what? We never went back to a curfew situation. For the most part, he'd come home far earlier than any curfew I would have set, that's how determined he was to prove to me that he was responsible. I only had to reprimand him once in the two years before he left for college, and on that occasion he offered up some sound reasoning. He was at a friend's house with a bunch of his football teammates until two or three o'clock in the morning, watching old game tapes. I knew where he was. He had called earlier in the evening. I knew what he was doing. I knew who was there with him. It wasn't as if he was out roaming the streets or bouncing around from one party to the next. His friend's parents were at home, and the kids kind of lost track of time. Still, he allowed that he should have called again, that he hadn't meant to worry me, and that it wouldn't happen again, so we kept to our no-curfew plan.

My younger son, Chris, is a different kind of kid—responsible and self-reliant but more of a free spirit than his brother—and for now we're sticking to a curfew. He doesn't want one and points to the lifting of Charles's curfew as the standard, but I'm not prepared to bend just yet. Why? Well, there are a lot of reasons, but I suppose the most compelling one is that I don't think he would have come to me with such a mature proposal if his brother hadn't set the precedent. Think of it: To take the initiative in an area such as this, to cloak yourself in the kind of constant good judgment such a request requires, is far different from simply wanting to live by the same rules as your older brother. Really, the biggest argument in favor of Charles and his no-curfew appeal was that he had thought to ask and that he had anticipated my concerns and answered them before doing so. The lesson here is that our parenting styles must remain fluid. We must be prepared to adapt to different situations, different kids, different environments, but underneath that willingness to adapt we must keep vigilant and focused and on point. The one piece that's not up for review in all this is the clarity of the message, the consistency, because without this firm foundation our kids will flounder. They just will.

A footnote: An important piece of this curfew dilemma will be moot in Chris's case because of a wonderful new law on the books in Georgia. Now drivers under eighteen can't be on the road after midnight. It's part of a statewide push to cut down on the rash of teenage drinking and driving fatalities, and I'm all for it. Another great piece to the law is that it limits the number of passengers under-eighteen drivers can carry in their cars to three, which still allows for double-dating and such but cuts down on the crazy pile-in mind-set that causes a lot of the reckless driving. Happily, as more and more states adopt these tougher guidelines on young drivers with more and more success, more and more states will follow suit—and, in many important respects, more and more parents will adjust their own expectations to mirror the changes in the laws.

Indulge me for a few more paragraphs on another Charles story, this one to illustrate how hard it can sometimes be for a parent—namely, me!—to stay on point, especially when your heart tells you to play a situation one way and your head another. Charles was ten years old, with a brand-new bicycle. At the time, we lived in a community near a park where a bicycle was almost a necessity for a small boy. The neighborhood kids were on their bikes all day long, back and forth to the park and to one another's houses, and Charles was in the bad habit of leaving his new bicycle outside, unwatched and unlocked, whenever he came into the house. Our house happened to be on a cut-through street to the park, there was a lot of traffic passing by our front yard, and it would have been nothing for some thief to eye the bicycle, pull over, and throw it into his car. I kept telling Charles his bicycle would get stolen and I wouldn't buy him another, but he wouldn't listen. Or he *couldn't* listen. Remember, he was just ten, which I guess is an age when you sometimes have to figure things out the hard way.

Sure enough, his bike was stolen one afternoon when he came in for lunch, and Charles was just despondent. All his friends had their great new bikes, and they were tooling around all day long, all summer long, and Charles had to keep up on foot. My heart plain broke for him, but without saying "I told you so," I felt it was important to keep the message clear.

"You'll have to figure it out," I said to him, reaching for that elusive parental mix of gentle and firm, "but one thing's for certain. I'm not buying you another one. That's what I've been telling you all along."

Well, Charles became determined to earn the money to buy another bicycle. He went from crestfallen to focused in no time flat. He made himself available for extra chores around the house, for a negotiated price. He knocked on doors, looking for odd jobs he could do for our neighbors. He told his grandparents and other relatives he didn't want presents for his birthday or for Christmas but cash instead. He went without a lot of the things he would have normally paid for himself, like an afternoon at the movies with his friends or a neat new video game, and he hoarded his meager allowance. It was difficult for me to stand by and watch him struggle like that. It would have been far easier for me to take him down to the store and drive home with a new bicycle in tow, or even to advance him the money so he could get back on his bicycle straightaway, but I didn't think that was what the situation called for, so I stuck to the plan.

It took a while, but he scraped together the money eventually, and I still remember the day we went down to the shop to get his new bicycle. Charles had one all picked out—and he had the money all counted out too. He reached into his own little pocket and held the money out to the clerk in the store, and he was all proud and grown-up and accomplished about it. He hadn't figured on the tax, though, so I sprang for the last twenty dollars or so behind his back. I didn't want to diminish his sense of pride, and at the same time I didn't want him to have to go home empty-handed, so I slipped the money to the clerk without Charles knowing, and he took that bicycle home feeling as if he'd earned it. And he had.

Throughout those long weeks of his saving, the temptation was to just give Charles the money, or have the bicycle waiting for him in the garage when he woke up one Saturday morning, but I had to be firm on it, and in retrospect I'm glad that I was. Charles too will point to this business with the stolen bicycle as one of the most important lessons of his growing up. It made him appreciate how much things cost. It made him take responsibility for his possessions as well as for his actions. It made him think ahead and think things through. In the end, he got the bike, and as a bonus he got all these important lessons, and he certainly never left his new wheels unattended in our front yard ever again.

It's really such a simple thing, when you break it down, for parents to keep clear and focused and on point in their directives and deliberations with their children. It's all about consistency, and setting a good example, and sticking to an overall plan, even when an easier path presents itself. And yet, day after day in my courtroom, I'd see parents who weren't clear with their children, parents who were visibly or outspokenly upset with their children for not meeting their expectations even though they'd never really articulated what those expectations were, parents who dropped the ball in such a way that their kids couldn't help but fumble it as well. Did these parents come right out and say, "Don't take the car without my permission"? No, of course not, but that much should have probably been clear. The miscommunication—or better, the noncommunication—came in the details. The drinking and driving. The staying out late. The hanging with the suspect crowd. The persistent lack of good judgment. The B-minus instead of the B-plus. What would happen, typically, is that the parent would get upset with the child, and eventually that anger would fade, with no clear communication or discussion about expectations. Soon there'd be a new disappointment to take its place, and what the kid would take in over time was that they could get away with this or that behavior. Their parents might rail or kick up some dust, but ultimately they'd let the matter slide. Or they'd let a whole bunch of small matters slide and draw the line on a much bigger deal, but how could they expect the child to pick up on those kinds of mixed signals? The lines have to be clear.

Naturally, if a kid gets away with something a half-dozen times, he's going to push his luck on the seventh occasion. Kids are designed to test the limits; it's in their nature. It's how I was as a kid. (If you're honest, it's probably how you were too.) It's almost like a challenge. You see how much stuff you can get away with, and you keep at it until you get nailed. And some parents are hardwired to look the other way, don't you think? It's in our nature to pray for the best. We close our eyes and try not to worry and hope a little thing will take care of itself, but when those little things start piling up, before we know it we're looking at a whole mountain of mess, and at the bottom of that mountain is our hopeless confusion.

Conflicting messages abound when it comes to our children. Our drug laws, for example, are a mess of inconsistency, most especially at the juvenile court level, and street-smart kids realize that even where the law is clear, it often doesn't get enforced. Wealthy kids know that if their parents are well connected, or if they can afford a top attorney, they probably won't face any serious consequences beyond a slap on the wrist. Plus, there's a kind of curbside justice in a lot of our communities, where if a police officer knows a child's mother or father he might take the kid home for a talking-to instead of hauling him into juvenile court. There is tremendous inequity in the ways our laws are enforced, and the net effect is that our kids are bewildered, confused by the mixed messages that are ultimately unclear. In Fulton County, for example, a kid who comes in on the misdemeanor of possessing a small amount of marijuana faces no real consequences; our system is so jammed up that there are no resources to process these kinds of cases according to the letter of the law. Yet in some of our smaller counties, where they don't have the same caseload, that same kid might face a stiff punishment.

I'm all for the zero-tolerance policies school districts have put in place to curb smoking, or fighting, or cheating.

If school administrators are clear on it, then the students will be clear on it too, and if the administration sacrifices any of that clarity by letting one student off the hook because he's able to talk his way out of a situation, or because his parents are powerful or savvy enough to rattle some cages, then everyone loses. Call me old-fashioned, but I'm one of those folks who happens to believe it's the kid who skates on his justified punishment who loses most of all. After all, over time, the messages that hit home to a kid like this are that rules are made to be broken and punishments set in stone can be washed away with pluck and swagger, and there's no place to head for from there but trouble.

Parents need to understand that there have got to be some serious consequences at home, regardless of the legal consequences or the consequences at school, whatever the transgression. As a judge, I was much more inclined to release a child to a parent who appeared on top of the situation than to a parent who thought the matter resolved with my disposition. Tell me you've already grounded your child, or taken away his car keys, or "volunteered" him for community service work at your church, or enrolled him in summer school. Tell me you understand the seriousness of the situation and you've got it under control, and it becomes a lot easier to find a hopeful, positive solution. There are a whole bunch of areas where the court shouldn't even become involved, and wouldn't, if parents would just step up and take some responsibility. Truancy matters, for example, are more successfully resolved at home than they ever are in court, because no court-appointed probation officer can be more aware of a school situation than a caring and involved parent. The best initiatives to forestall truancy, including our Truancy Intervention Project, have called for a considerable degree of parental cooperation.

I've been at this judge thing a good long time by this point, but I've been at this parent thing a lot longer, and the two roles are more connected than a lot of folks think. As a parent, and as a judge, I try to talk in terms children can understand so that my message is abundantly clear. One of my favorite courtroom metaphors for getting kids to take more responsibility for their actions grew out of a conversation I had with my own children.

"Understand football?" I'll say, and from the teenage boys before my bench I'll usually get back a "Yes, ma'am," or some such.

"Understand a quarterback's role?" I'll try.

"Yes, ma'am," I'll get back again.

"Good," I'll say, "cause my rules are slightly different. Here's how you play my game. You're the quarterback. You've got the ball and you've got to move it forward. The only thing is, you can't pass it. You can't punt it. You can't lateral it. You can't even fumble it. So what is it there's left for you to do?"

They'll usually think about this for a bit before answering sheepishly, "I've got to run with it, Judge."

"That's right," I'll say. "You can't do anything but run with it. Nobody else can handle this ball but you.

This is where you are, and that right there, off in the distance, where you want to be, that's the goal line. The goal line is where your dreams are. And your parents can be there for you on the sidelines, cheering you on. They're like the coaches. And there are other folks to help you too. The probation officers are going to round out that coaching team and in some cases maybe a caseworker or a social worker. Wherever you look, there will be people cheering you on, but it's on you. You've got the ball, and you've got to run with it, and you've got to get to the goal line. In my court, those are the rules."

And I'm here to tell you that they get it. They really do. There's a way to get that message home to every child, no matter how hard-core, no matter how far gone. I'm not a perfect parent, and I'm not a perfect judge, but if I strive to keep clear and focused with *all* my children, they'll get my point and stay with the program.

Over the years I've developed a keen sense of radar for when a child is telling it to me straight—what some folks might call a BS detector. Kids have that same ability, only with them it's innate. They can see an adult giving them a line a mile away; there's nothing to develop.

In my case, when that radar goes off, I'll give that child another chance to come clean.

*Judge, I've been running with the wrong crowd. . . .*

I can't tell you how many times I've heard that line served up in court—as a defense, no less! It's right up there on the all-time-greatest-hits list of unbelievable excuses kids come up with in my courtroom. And each time, I've answered back, "Well, if I could just lock up the wrong crowd, I could solve the major crime problems in this city!"

*Judge, I was just in the wrong place at the wrong time. . . .*

That's another good one, and underneath these knee-jerk excuses for variously bad behavior there is always a set of dangerously mixed messages at home—or worse, no messages at all.

Put it back on the child, though, and together you can see your way clear to resolution. Put that ball back in his hands and get him to run with it. The wrong crowd? I wouldn't buy it: *Are you telling me that you're stupid enough to be the lookout when your buddies break into someone's house, just because they told you to? You've got to start thinking for yourself. These no-good, jive-ass, trifling folk that got you into trouble? They're not around to bail you out, are they? They're not here with you in court today, are they? A true friend will never lead you to no good. A true friend will never lead you into harm's way. You've got the kinds of friends who ask you to hold the money in a drug deal so there won't be any money on them when the cops come round. You've got the kinds of friends who ask you to hold their pinched merchandise on a shoplifting spree at the mall. Look at how quickly they run in the other direction when the security officer turns up and you're left holding the bag. Or look how quickly they point to you when they get pulled over driving a hot car. Where are your friends now?*

You'd be amazed how many kids have never been empowered to take charge of their own circumstances in a positive way. How many kids have never been told, "I believe you can do this." How many kids have never been told, "I'm pulling for you." The message must be clear and consistent, whatever it happens to be. Your dreams for your child. Your child's dreams for himself. Your expectations. Your rules. Set it all down, in clear, simple terms, and hold each other to it—for it is in the holding that we keep true to ourselves and to each other.

## *A Place to Belong*

I never thought I'd be a judge.

A lawyer, yes, but just to keep my options open.

Realize, we weren't rich when I was a child, not in monetary terms, but we were wealthy in many other ways. In every important respect, we had everything. Love. Faith. Harmony. Hope. This last was essential—indeed, it was perhaps the key difference between growing up in my generation and growing up in my parents' or grandparents'. We were allowed to attach real prospects to our dreams.

My two younger brothers, Paul and Kolen, and I were enormously blessed with wonderful parents who taught us to trust in ourselves, in one another, and in our circumstances. One of the great gifts from my parents was that I grew up never believing that being black or female was in any way problematic. They never gave me that frame of reference. The message to me was that my race and my gender were not curses but blessings; they were the reasons why I would soar, not excuses for why I might be grounded. I could do anything, be anything, pursue anything. I wasn't born into this world with two strikes against me but with two big points in my favor, and I walked with confidence into situations another black girl might have avoided.

From time to time, however, that confidence was shaken. I still carry very vivid memories of being knocked from a water fountain step stool in a Sears Roebuck in Florence, South Carolina, and not understanding why. I had just turned five. I was visiting my grandmother, and we were on an outing. I could read at a very early age, and the sign cautioning Whites Only struck me as arbitrary and curious. I wondered how it applied to me and to this little water fountain. I wondered if the water tasted different in the white fountain than it did in mine, so when my grandmother turned her back for a moment I rushed to the white fountain to see for myself.

I don't know if I ever got that taste. All I remember now is that I got knocked down by two white boys (one of whom was exceedingly chubby, with bright red hair!) and that I made my grandmother angry. I realize now that she feared for my safety. I could have been dragged out of there; she might not have seen me again. But as a child, all I knew was that I had been knocked down, pushed away from something I wanted. I didn't like it.

All of which, in a fundamental way, takes me back to keeping my options open, to the way I was raised, to the faith I was taught to have in myself. I don't mean to suggest that I walked through life with a bullheaded, scheming, win-at-all-costs personality, but at the same time I was never the sort to be denied easily, and this was especially so as I looked to chart a career. The thing to do, as a young lawyer graduating from Emory Law School, was to reach for the biggest door, the one that would lead to the widest range of opportunity—to one, perhaps, that had never been opened before. For me, I determined that door would be a coveted clerkship in federal district court. Actually, it wasn't yet a federal clerkship, and it wasn't yet coveted by too many folks other than me, but I looked to pick my spots in places others had yet to consider.

Again, I never meant to be a judge, never even meant to consider it, but right out of the gate I meant to work for one—specifically, for Judge Horace Ward, the noted civil rights activist, attorney, and one-time state senator who was up for appointment to the federal bench under President Jimmy Carter, another one of Georgia's own. Judge Ward had done his political coming of age around the same time as President Carter, around the same places. It was clear to almost everyone who knew about such things that the president would tap the judge to fill the first appropriate vacancy on the federal bench, thereby making Horace Ward the first black federal judge in the state of Georgia—indeed, in the Deep South. That appointment might be a year or two in coming, but it was clearly coming, and in the meantime I wanted to work for this great, proud man, a man who had spent his early career breaking down barriers so that some upstart like me might come bursting through.

I was looking to cut my own road, and as far as I knew that road ran alongside Judge Ward. Some of the people I trusted were trying to talk me out of this view, convincing me to go to a big law firm, or to the district attorney's office, or to a corporate counsel position at one of Atlanta's top companies, but I was unwavering. These folks told me I was foolish to forego those big salaries and certain career paths, but at that time in my life I didn't care about money or making partner or any of those things. I knew that if I could convince Judge Ward to take me on now, before his federal appointment,

I'd have the inside track for a spot as a federal district clerk, and I knew that when the time came, hundreds of young attorneys throughout the country would be vying for the same job. And I knew that that post, shot through with prestige and purpose, would open almost every door going forward.

And that's just how it happened. When President Carter's appointment ultimately came down, people were willing to leave partner-track positions at top law firms (and at top salaries); people were willing to leave the U.S. Attorney's office; people were willing to do whatever it took to throw in with Judge Ward, but I was already on board. I'd convinced Judge Ward to take me on as a clerk in federal court, and I had worked tremendously hard to do a good, thorough job. My counterparts would be buried in the basements of law libraries, never seeing the inside of a courtroom, and I would be sitting as a clerk on major federal cases, drafting opinions that would be appealed to the appellate court and the U.S. Supreme Court.

And yet for all my forethought and consideration, I was in no way prepared for the awesome responsibility and the overwhelming sense of history and moment that accompanied the federal job. Clearly, it was Judge Ward's moment, and his place in history, but it was also mine—once removed. Realize, when Judge Ward was sworn in to the federal bench, he took the oath of office in the same ornate federal courtroom where he had been denied admission to the University of Georgia Law School nearly a quarter-century before. Back in 1950, just out of Morehouse College, Horace Ward had become a kind of national test case, tapped by the NAACP and its legal defense fund to spur integration of the top white schools in the South. Thurgood Marshall was assigned as his lawyer—this back before Ward himself famously worked to integrate the undergraduate school at the University of Georgia on behalf of Hamilton Holmes (my next door neighbor when I was a child!) and Charlayne Hunter. The case dragged on for six years, at the other end of which Horace Ward enrolled in law school at Northwestern University and vowed someday to return to Georgia, on his own terms.

I sat there, in the jury box, in the very courtroom where Thurgood Marshall had unsuccessfully argued for the admission of this fine and formidable man, and I was quietly overcome. I was twenty-six years old, and the direct line from the girl I had been at five, struggling to sip from the Whites Only fountain at a Sears Roebuck in Florence, South Carolina, to the woman I was becoming, the first black federal district clerk, male or female, in the state's colorful history, was everywhere apparent.

Horace Ward would not be denied a second time in this courtroom, and I would never again be pushed from any water fountains. I would get my taste, after all. I would realize my dreams. I still never thought I'd be a judge—I didn't think I had the temperament for it—but for the time being I was content to work at the heels of the best judge I knew, a dignified, dedicated man, making history in the same room where opportunity and equality had once been denied.

## Simple Strategy #3

# Listen Carefully

There's a haunting image from my first week on the bench that I've never been able to forget, and it finds its way into these pages for the way it forced me to actively reconsider that difficult crawlspace between career and family. It also hit home, hard, as I imagine it would have done in your home as well.

I was brand-new enough to the juvenile court system that the slightest jolt to my sense of humanity and fairness could send me reeling. Indeed, I never got used to the thought of a child done wrong, and I never will. Show me an injustice visited on a helpless child, and I'll be shaken—especially in those first weeks on the job. And that's just how it started: A little boy, who couldn't have been more than five or six years old, stood downstairs in our processing area one afternoon looking completely disoriented. He was dehydrated and numb to the whir of activity of these well-meaning strangers all around. He hardly moved; his face hardly registered his circumstance; he might have been staring blankly at a cartoon on television. He had been picked up on the median of North Side Drive, a very busy thoroughfare, three lanes in each direction, where he had been standing stock-still, frozen in the middle of all that traffic, unable to tell the police officer who found him how he'd gotten to just that place at just that time. It was the most upsetting thing, to think of this innocent child out there on that road. What a harrowing, heartbreaking picture! Really, he could have been killed. As it was, he must have been scared half to death.

It wasn't even my case, but it rattled me just the same. I was downstairs when they brought this child in, working my way up and down the halls, trying to learn everybody's name in this strange, bustling new place where I'd decided to remake my career, and I was just floored by this sad, vulnerable little boy. He was brought in so that we could issue an emergency order of protection, to take him into temporary custody and arrange for appropriate medical care, and I was struck by the fact that the child couldn't tell us his last name, his phone number, or his address. For my new colleagues, this was routine; for me, this was so far off the map of my everyday experience that I could have asked for directions and still turned up lost. The little boy wouldn't say a word, and into his silence I imagined every worst-case scenario, every sensational headline, every textbook *don't* from every expert who'd ever worked with desperate children. We learned later that the boy's mother was a drug addict, that she'd lost track of him while she was off somewhere getting high, and that the child and his mother had been living a very transient life. It's a wonder they'd sailed under our radar for as long as they had—although at the same time it was no surprise at all. There must have been hundreds of broken families, perhaps thousands, slipping through the cracks in our system in much the same way all across Fulton County. I was still new enough to the process to be surprised, and thrown, to see this little boy as a kind of poster child for the much larger problem.

I went home that night haunted by this child. I closed my eyes, and there he was. I looked away, and there he was. I can still see his bewildered face all these years later, but that first night it was with me most of all, and the reason it hit me so hard, I think, was that I had children close to the same age. Charles was about nine and a half years old at the time, and Chris was just five, and I looked into the eyes of my own children—happy, healthy, heart strong—and saw this little boy, how his life might have turned out under different circumstances, how my own children might have fared under the same raw deal.

Try as I might, I couldn't shake this little boy from my thoughts. He was everywhere and all around. Remember, I was still new to the job, and I had yet to develop the survival skills I'd need to get by in one emotional piece. What I knew in theory I did not yet realize firsthand—chiefly, that it's a tough thing to have your heart broken a dozen times each day and to go home each night and find ways to patch it back together and fill it back up. I had my own family to worry about, my own hopes and dreams that needed nurturing, and I knew enough to recognize that these would require my full attention when I took the robes off at the end of each session. You can't just listen with one ear when your kids tell you about their day, and you can't focus on the tasks before you at work if you're worried about what's going on at home. It's the same old song, the dilemma of every working mother of my generation—how to shut off those professional valves to let our personal lives flow freely—but I was a rookie judge and making this part up as I went along, and it took this one little boy to set me right. I'd encountered this child after I had finished the last case on my afternoon calendar, and for some inexplicable reason I became totally preoccupied with him for the entire drive home, to the point where when I finally found myself in my kitchen, getting dinner ready, I realized I wasn't at all focused on what was going on in my own household.

Now I know it doesn't sound all that bad to a casual onlooker, to be somewhat tuned out to what's going on right in front of us. We've all been distracted by work in such a way that it colors our time with our children, right? But to me just then it was a great big deal—actually, it still is—and I was quite upset by it. I'll never lose that moment. I can even remember what I was making for dinner that night—broiled steaks, with broccoli and canned yams.

Chris and Charles were in the kitchen, doing their usual thing, homework and whatnot, and I went through the motions of asking about their days. For a while I actually listened, but after another while Chris crossed over to me and started tugging on my skirt.

"Mommy, Mommy," he said, desperate to pull me back to whatever it was he was saying, "you aren't listening to me!" He was frustrated, confused, and completely put off by my inattention.

*Mommy, Mommy, you aren't listening to me!*

It struck me like a bad dream. Chris was right. I was back in that processing area, fixated on that helpless little boy. My heart was aching so for that child that there was no room in it for my own two children, and I thought to myself, This is not good. Hell, it was more than *not good;* it was downright disturbing. I can't stress how upset I was by my level of distraction. It's not as if I was new to the workplace and was simply having some trouble readjusting the balance between career and family. No, I'd been a working mother as long as I'd been a mother, and nothing like this had ever happened in the whole of my professional experience. What had changed in the past week was the kind of work I was doing, and it took this tug-on-my-skirt moment with Charles and Chris to get me to realize that this was no ordinary job, presiding in juvenile court, looking face first at the hopeless and desperate and dispirited youth of Atlanta's broken families, and that if I hoped to succeed as a judge and as a mother I'd need to develop some tools to keep the one role from bleeding into the other and both from bleeding me dry.

I suppose every parent has a moment like this, best intentions aside. Whether it's work, or an argument with a spouse, or worries over money or relatives, or some other pressing issue, there's usually something to pull our careful, focused attention from our children. For me, this was the first time I wasn't able to engage with my kids when I knew full well that I needed to engage with my kids, and I didn't like the way it left me feeling. I didn't like what might fall from this one piece of inattention, so I vowed it wouldn't happen again.

I set this out as a warning, as it came to me—a not-so-gentle reminder that as parents we must listen carefully to whatever it is our children have to tell us, in whatever way they choose to communicate. The great buzzword here is *carefully*. We must listen with great care, and we must listen fully, on many different levels, at all times. Children know when you're tuned in to them and when you're tuned out. Absolutely. You can't simply go through the motions and phone this parent-child thing in at the end of the day. You can't ask questions and not listen to the answers. You can't be there in a physical sense without being there in every other sense as well—emotionally, wholeheartedly, spiritually, and every other which way besides. If you carve out some time to spend with your child, then spend that time in a quality way. (And spend it gladly!) If you ask your child how his day was, then be prepared to listen to his answer. Fully. And while you're at it, be prepared to share some age-appropriate details from your own day as well.

This wasn't news to me, that night in my kitchen, with Chris tugging at my dress in frustration. I had it in my head from the moment my kids could communicate that this "How was your day?" time was all-important. I made it part of our routine, on the theory that if you get your children in the habit of communicating openly when they're young, then hopefully they won't shut down when they get older, when it becomes even more critical to understand what's going on in their lives and for them to understand that they can come to you with anything. After all, how much can go wrong in kindergarten, in a Montessori school? Not a whole lot, right? But that wasn't the point. The point was instilling in my children the good habit of downloading the stuff of their days to a concerned, loving, carefully attentive parent at the end of each day, without censure or fear of reprisal.

So that's what we did in our house, every day, without fail. Sometimes I'd have to bite my tongue and hear my kids out before doubling back and saying, "Okay, let's talk some more about that fight on the playground," or whatever it was that caught my attention during the initial report, because I discovered early on that if I cut my kids off or redirected them in any way, I'd never get the rest of the story. But not once, in the six or seven years we'd been doing this, since Charles was old enough to speak in full sentences, had I ever been so distracted by the stuff of my own days that I zoned out the way I did this one distressing night during my first week as a judge.

*Mommy, Mommy, you aren't listening to me!*

No, I wasn't, not a lick, and I was plainly bothered by this. I couldn't even grab a scrap of what Chris had said, a little detail that would have given me the gist and allowed me to follow along until I could get up to speed. Some kid's name. Some general location or time-of-day stamp to pin on the story. I had nothing, and Chris could see I had nothing, and it scared me because I realized how easily I could have fallen down the slippery slope that propels parents inexorably away from their children. It must have scared Chris too, in what ways he could understand, judging by the way he was tugging so frantically at my dress. He'd never seen his mother like this, and he didn't like it. I didn't like it either, and I knew it would be downhill from there if I didn't do something about it. And quick.

It was a critical lesson. Obvious, perhaps, but no less critical. I couldn't bring the courtroom home with me. No matter what I had seen. No matter what I had heard. No matter what I was feeling. I'd only been on the job a week or so, but I'd seen and heard and felt enough to know I was about to see and hear and feel far worse and that there would be no end to it. Yet I also knew that I couldn't simply flip a switch and be done with the pressures and heartbreaks of each and every case. So I began to put myself through a decompression ritual on the drive home each afternoon. That time in the car became my time to calm down, relax, vent . . . whatever I needed to get to that next, all-important part of my day, that quality time with Charles and Chris. I gave myself the space to think through everything I needed to think through from court, to run stuff back and forth in my mind until I'd run out of thinking. Basically, I gave myself time to unwind before walking through my front door. There were some nights when I wasn't quite through with the unwinding when I pulled into the drive, so I stayed in the car for a few moments until I could focus on what I would find on the other side of the front door. In the beginning I had the ritual down to the point where the turn of the key in the front door lock was the dividing line. Once I put that key in the lock, I was no longer a judge. I was a mother, a wife, a daughter, a sister, a friend. . . .

The power was in the ritual, if that makes any sense. It came from the routine, from going through these simple motions until they became second nature. Were there days when the tragedies and misfortunes of some of my juvenile court families seeped into my own household? I'm not going to lie. Yes, absolutely, no question about it, I'm only human. I did the best I could, and in the beginning it wasn't always good enough. Let's face it, there is no foolproof, fail-safe system to keep all outside influences from hearth and home. But I learned to compartmentalize that part of my life, to leave it at the door and pick it back up again the next morning for the drive back to the courthouse. I got better at it as I went along.

Yes, it took a real effort, and a real long time, but I got there eventually, and along the way I refused to be beaten down by the affecting stories that found me on the bench. Abused children. Neglected children. Abandoned children. Kids gone bad for reasons of their own making. There was a fifteen-year-old paid assassin, a fifteen-year-old prostitute whose own father was her pimp, and another fifteen-year-old who'd given birth to her fourth child. I'd seen kids starved to death by their crack-addicted mothers who'd simply forgotten to feed them. I'd seen a little girl locked in a closet for over a year and a little boy prostituted by his parents in exchange for a fix. Every time I thought I'd seen it all, there was something new and horrifying. Each case added to the emotional weight of the ones before, but this simple key-in-the-lock ritual allowed me to set the weight aside, to be something other than a juvenile court judge . . . at least until the next case came around on my calendar. If I caught myself drifting, I dragged myself back into the moment with my kids. Or I put things on pause for a couple of beats, excused myself to the bathroom if I had to, so that I could splash some water on my face and refocus. Whatever it took—and soon enough, all it took was that conscious effort to decompress on the drive home, the keen and constant willingness to keep the professional parts of my life away from the personal parts.

Scratch the surface of almost every case I heard in juvenile court and you'd find a parent who couldn't take the time to listen to his or her child. To listen *carefully*. To focus. To keep an open mind behind an open door—with open eyes and an open ear. I'm not much for beating home a point, but the openness here is key, because for every parent who was unavailable to a child, for every instance of close-mindedness, for every turn of a blind eye or a deaf ear, there was a kid in trouble. The cause and effect was transparent, and the trouble ran from bad to worse.

A teenage girl with good grades and no prior record. Both parents were high-achieving, successful individuals, prominent in their community. Both parents, if asked, would have spun deeply held dreams for their daughter that included a good college, a promising career, and a family of her own. And for the most part their daughter kept up her end of the deal; she was involved in all kinds of extracurricular activities, was an honor roll student, and was by all outward appearances headed down the right road. Until one night when she sneaked out of her house, hopped into the family Volvo, and went joyriding right smack into another car. No one was hurt, mercifully, but there was enough property damage to the other vehicle that the case bounced upstairs to my calendar, and when I started talking to this young girl, the pieces didn't seem to add up. Here was this bright, hereditarily high-achieving child who had never wanted for anything in any kind of material sense, who was doing well in school, who seemed to have the *right* circle of friends. And yet she was ducking out of her house in the middle of the night, up to no good for no good reason. Now if you want to take a hard line, you might suggest that there's never a good reason to be up to no good but usually there's an explanation. Usually there are other kids involved. Peer pressure. Drugs or alcohol. The promise of cheap thrills and a story to tell. But this girl was acting on her own, clean and sober, grabbing her father's keys and taking his car out for a spin.

Doesn't make any sense, right?

As it turned out, this wasn't the first time the girl had taken the family car without permission. She was only fourteen or fifteen, under legal driving age, but apparently she made a habit of these after-hours drives. This was the first time she'd gotten into an accident, however, and on paper it struck me as a classic case of an otherwise good kid acting out to get attention. During our hearing I kept looking at this child and then at her parents, back and forth, wondering how we had gotten to this place. How did this teenager go from being a good kid to standing before me in my courtroom facing these kinds of charges?

Well, I didn't have to look too deeply to begin to see the answer. The girl's father couldn't have been less interested in what was going on. He tried to say the right things when called upon to be supportive of his daughter, but he offered up knee-jerk, cookie-cutter responses; everything about his body language told me he would rather have been anyplace in the world but in my courtroom at just that moment. His heart wasn't in it. His head was someplace else. He couldn't be bothered. He had this attitude that said to his child, "Look, I've done all these things for you, I've provided you with all these things, and I will not accept this kind of behavior in return." He didn't want to hear that his daughter so desperately needed his attention that she cracked up the family car to get it. That wasn't part of his bargain. His deal was to pull on his powerful strings, to dig deep, to write whatever checks he needed to write to keep his daughter in the best clothes, the best schools, the best after-school programs, and he expected these things to compensate for his time and attention.

Of course they could not, and the more I spoke with this child and her parents, the more I felt the huge disconnect between them. They were good people, I felt sure. The father was a bit domineering, and the mother was a bit too submissive, and together they didn't have a clue, but they had no intention of winding up in my courtroom. Clearly, they didn't *want* their daughter in this kind of trouble, and they didn't exactly neglect her in the classic sense of the term, but at the same time they hadn't been prepared to invest their full attention in her emotional development. They couldn't be there for their child so much as they could hire someone else to be there for her. They could have seen this trouble coming if they'd thought to look. At times during the tense back-and-forth in that courtroom, I got the feeling that this child was somehow in the way—at least that's how she was made to feel by her parents. And she said as much, in what ways she could, to the point where she had to crash her dad's Volvo into another car to get his attention.

How many times have we been caught in the too-easy trap of pacifying our kids with a toy or a video or another distraction when what was really called for was our careful attention? Let's be honest with ourselves and admit that it happens to the best of us. We come home from work, we're tired, we've got a headache in need of a vat of aspirin . . . the last thing we need is to deal with our kids' squabbling or to run interference for some petty problem with friends at school that our distracted minds feel could certainly wait another day or two. But guess what? String too many of these too-easy traps together and you're left with the kind of minefield that was facing this high-achieving young girl, one she couldn't get past without this over-the-top cry for attention.

Kids tug on our skirts (and pants!) in all kinds of ways, and it falls to us to read the signs. We can't expect our children to articulate what they're feeling on this score; if they can, that's great, but we can't *expect* it. What we can expect is that if we don't make the time to listen to our children, if we don't make ourselves fully available to them, if we give as little of our time as possible instead of as much as we can afford, then the odds of our children making bad choices for bad reasons increase exponentially. Of course we can do all these things, and everything else besides, and still find ourselves in someone's juvenile court somewhere, but as good parents we've got to play the odds, right? They're all we've got. And the odds tell us that if all of a sudden you've got a kid who doesn't want to come home, something's wrong. If all of a sudden you've got a struggling report card in place of a striving one, something's wrong. If all of a sudden your bright, outgoing child has turned sullen and introspective, something's wrong. If all of a sudden an open, honest line of communication becomes closed and deceitful, something's wrong.

Sometimes we can anticipate these changes. Better, we must. One of the last cases I heard before I left the bench struck me as one of those stories you long to put on rewind, to find some magical power to go back in time and get it right. Another young girl crying out for help, in a manner even more desperate and destructive than the Volvo-stealing teenager. And at the bottom there was an inattentive parent who could have solved the whole problem before it even materialized. This child was caught in a regrettable custody arrangement. Her parents had split when she was about ten, her father remarried when she was about twelve, and she ended up living with her father and his new wife. At the time of the custody hearing, the girl was an A student with no indication of any trouble to come—and yet the trouble didn't even bother to knock. Almost immediately the girl started sneaking out of her house at night, running with a suspect crowd, ditching school, abusing drugs. Her father worked long hours and traveled frequently, and the girl was left with a stepmother who didn't appear to be all that interested in any kind of parenting role. In what ways she could, the girl told her father how miserable she was to be left behind with this woman with whom she had no real connection. She was lonely, with no sense of family, no sense of belonging, and if she tried to talk to her father about it, he'd tell her to be patient, that things would get better, that he'd have more time to spend at home once he got done with this or that deal.

But nothing changed, and the girl's behavior went

south in a reckless, headlong kind of way. Either the father couldn't hear what his daughter was telling him or he didn't want to hear it, but I can't imagine that he missed the signs. By high school, the girl's grades were tanking. The stepmother never knew where she was or what she was doing. She was truant from school and running away from home and taking all kinds of drugs and sleeping around. It was such a sad, sad shame. It's always a shame, but this struck me as especially sad. This was a beautiful child who had come from opportunity. She was bright. And she was trying. I really believe that at various points early on in these troubles she was the only one in her blended family looking for a hopeful solution. She kept asking her father to spend more time with her. She kept coming up with ideas for things they might do together. She was quite clear about how unhappy she was. She was entirely reasonable about her unreasonable situation. For a time.

In the end, the girl popped up on my calendar as a teenage runaway with a growing history of drug-related and criminal activity, and we slogged through the broken family dynamic trying to find a way to keep her at home with some intensive family counseling. Even here, at this last resort, I could see how tuned out the father was to all the trouble. It was a constant screaming match in my courtroom. The daughter would say one thing, and the father would deny it. She'd look one way, and he'd look the other. His instinct was to reject her claims as invalid, to counter every negative with a concocted positive, to turn away from every opportunity to engage lovingly and purposefully with his child. The man just wouldn't listen, even at the crisis point of a juvenile court hearing for his daughter's truancy and related delinquent activity.

This was one of the last cases I heard before I left the bench, and it left me with a heavy heart. Here I thought I had done my part to stem the tide of these kinds of cases, and this girl throws away her opportunities, and lands face first in the kind of trouble nobody could fix. That was the tragic kicker to this tragic story, and it found me after I'd left the bench, when the caseworker informed me that the girl had tested HIV-positive. All that time on the street, running around, had finally issued the kind of sentence I'd been working to avoid in my courtroom. And the most chilling part of the whole deal was that this young girl's life could have played out in an entirely different way. Goodness, she was just a child! I thought to myself, If she had only gotten her father to listen. If I had only gotten her father to listen. If the man had grasped from the very beginning how important it was for an adolescent girl to have a strong adult presence in her life. Like I said, it's one of those stories you want to rewind so that it has a chance to come out another way, a better way.

I think back to that moment in my own kitchen with my own kids, all those years ago, and shudder at the memory. To many of you it might have been a nothing moment, a tossed-off frustration between a needy kid and a slightly harried parent, but to me it was an epiphany, a wake-up call, a not-so-gentle reminder that I had better start practicing what I was preaching. It was the difference between listening and not listening, between becoming the kind of mother I was suddenly afraid of and the kind of mother I wanted to be, between the desperate, destructive, helpless children I'd seen in my courtroom and the joyful, loving, supported children I saw in my home and in my dreams.

Listen. Fully and with great care. And everything else will fall into place behind your open ears, your open eyes, your open mind, and your open heart.

## *Answering the Charge*

Now even if I had planned on being a judge, I certainly didn't fit the profile. For one thing I didn't think I had the disposition for it. A judge has to sit still and keep calm, remain even-tempered and mind her manners, and folks who know me will line up to tell you that just isn't me. I've got too much to say to hold my tongue for too terribly long. In fact, I'll fess up here and admit that when I did finally ascend to the bench I had to sit on my hands for a couple of weeks just to keep myself from objecting! I'd wonder why an attorney didn't think to object or to pursue another line of questioning or why he or she didn't do this or that. I'd try other folks' cases for them in my head, because

I'd been bitten hard by the litigation bug, and underneath I sometimes thought I'd do a better job of it than some of these other attorneys. For the longest time, that's who I was.

When it came time to move on from federal district court, I took a job in the legal department at Delta Air Lines, another Atlanta institution, trying cases. In the beginning I tried mostly labor cases, but after a time I moved on to antitrust and contract cases too. It was the perfect job at the perfect time in my young career. I was married. We were starting a family. The money was good, the perks were welcome, and the hours weren't all that unreasonable. It seemed to me there had to be a way to combine the job of being an attorney with the job of being a mother, and I set about finding it—with some success, I should add. I worked hard, but I was close enough to home to break away for a few hours each night to deal with dinner and homework and bedtime and whatnot, and I could slip back to the office when necessary to tie up loose ends and prepare for the next morning.

Things went on in this way for six or seven years—good years mostly, although over time my marriage began to fall apart. I don't mean to get into the details of how and why my marriage eventually ended, because what passes between two people should remain between those two people, but from the marriage came two wonderful sons,

Charles and Chris. The point here, though, is that my job at Delta was a great fit for a working mother and that I was content to keep doing what I was doing as long as it needed doing. It may not have been the job of my dreams, but it became a dream job—it truly did! I was the highest-ranking woman of color at Delta, with a mentor who would become president and chief executive officer of the company.

And then the world changed. Two Delta planes crashed at the same Dallas airport in separate accidents less than two years apart, and I was dispatched as part of the crisis management team, dealing with the press, running interference for the families of the victims, doing what needed to be done. What a lot of people don't remember is that the very next day after the second crash, the very same flight with the very same flight number aborted on takeoff. No one was hurt, but the press was all over Delta, and at one point in all the resulting back-and-forth I looked up and realized that I was the one who was out in front deflecting all these charges against the company, reassuring the public that our skies were safe. Without ever really meaning to, I had become one of the public faces of a major airline at a time of profound and stressful crisis.

And apparently I was doing a good job of it, because the very next week, after some of the dust had cleared, I was called into my mentor's office and asked to join our press relations and crisis management effort full-time, on a permanent basis. I thought, Isn't this interesting? And it was. It happened to be a very compelling time during Delta's growth as an airline. The sudden shift from litigating cases presented an interesting new challenge—one I would never have considered even as I was happy to accept it. The company had just taken over some key Pan Am routes; we were beginning to stretch our muscles internationally; the industry as a whole was starting to pay attention to some of the concerns regarding terrorism and airline safety that are now at the forefront of our thinking. And there I was, with my great legal education and thoroughgoing background in litigation, plugging the holes in our public relations program. I didn't stop to think about what I was doing, or why I was doing it, or how I had come to the role; I just went out and did it in the best way I knew how, and what happened in the doing was that people started to pay attention. Delta couldn't help but make news, and it was often my face attached to the story, my name on the press release, my spin on the spin control.

I don't know that this attention-getting aspect of my new job was a good thing or a bad thing, but it was most definitely a thing. Something to be considered, at least, which was why when I took a call from our family friend Romae Turner Powell I didn't think much of it. Judge Powell was a tough-as-nails civil rights attorney who'd known my parents for years; she was the first black person in Georgia, male or female, appointed as a state-level judge. She presided over the Fulton County Juvenile Court system, which was, of course, a major accomplishment, particularly at that point in the state's history. She was also something of a mentor for me. She helped me decide to go to law school. She helped me realize, when she caught me sneaking out of an NAACP dinner to call home and check on my infant son, Charles, that my place was with my child and that there'd be a thousand other dinners. Every year or two we'd alight in each other's path and she'd redirect me in mine, so it didn't exactly surprise me when she called one Saturday afternoon just to talk. What was surprising was that she caught me alone and unencumbered; the kids were out with their dad, and the house was quiet. I had all the time in the world to talk, and we filled it up pretty quick. Judge Powell had never been much for chitchat, but here she was, talking. About my life. About my children. About my parents. About my dreams. She started asking me about my job, asking me if I was challenged by what I was doing, if I found it fulfilling. She asked if I had ever considered the kind of career whose rewards would be somewhat less tangible, where I might make more of a positive impact on other people's lives. She told me about her job on the bench, helping to repair the broken young lives of our community. She told me how the work sometimes left her feeling richer than she'd ever dreamed and at the same time more spent then she'd ever imagined.

We went back and forth in this way for over an hour, until Judge Powell got around to wondering what it might take to get me to leave Delta Air Lines. Frankly, that was the furthest thing from my mind. My goal was to distinguish myself in such a way that I would become an officer of the company, after which I might gracefully retire to a position on the board of directors. That was the plan, and I told Judge Powell as much.

"So there's no job you could imagine that would get you to leave?" she tried again, pressing her point.

"No," I said, "not that I can imagine."

Clearly, Judge Powell had a job in mind for me—her job. I didn't realize it at the time, but the judge was terminally ill with lung cancer—in fact, she kept coughing during our talk, which she attributed to a lingering bronchitis—and I learned later that she knew she was dying. We never talked about it beyond this one conversation, but I suppose Judge Powell was trying to line up someone with some of the same values and ideals she'd brought to the bench, and I further suppose that that someone was this brash young attorney who'd lately been in Atlanta's face as the ever-present spokesperson for Delta Air Lines. It all added up, after Judge Powell passed away about two months later, and I found myself still considering her charge.

Of course Judge Powell was in no position to appoint her successor. All she could do was push my buttons and get me to apply for the job. Typically, the governor makes a judicial appointment at the state level, but juvenile court judges are put through slightly different paces and appointed to a four-year term by a panel of superior court judges from the county.

I never did get to see Judge Powell after that lengthy phone conversation. She was hospitalized soon after, and she didn't want anyone to see her, but I took her counsel to heart. I really did. I've since joked that I must have been out of my mind to seriously consider leaving Delta Air Lines for a judicial appointment. I was making a very generous salary. I had a 401K plan. I had a mentor who'd just been named the president of the company and who was planning to take me along for the ride. I flew first class all over the world. I had litigated cases in federal courts all across the country, and yet I'd never even been inside a juvenile court. Why would I take something like that on? There are some folks who want so badly to be a judge, who want so badly to wear those robes that even a bathrobe starts to look good to them, but that wasn't me. I didn't care for the power or the authority. I didn't care for the robes or the other trappings of the bench.

What I cared about was doing good work, work that mattered, work that made a difference, and now that Judge Powell had started me thinking on it, I realized that my work in crisis management and media relations wouldn't change lives. Oh, it was important to Delta's bottom line and public image, and from time to time it was important to certain individuals, but if I weren't there to do it there'd be someone else in my place, and after a while even my supporters at the company might not know the difference. Sometimes the truth is tough to take, and this was one of those times. I struggled with it, because at the same time I'd grown extremely comfortable with the job. I liked the pay. I liked the benefits. I liked my colleagues. I liked the stability. I liked that we didn't climb into a station wagon for our family vacations but onto an airplane. The whole package was so different from the way I was raised that it was hard to see past it, even as I forced myself to look.

And yet underneath this comfort was Judge Powell's charge, and underneath the charge was the dawning realization that if I was ever going to step outside myself and make a positive difference with my life and a positive contribution to my community, I had better do it soon. I was thirty-nine years old. I hadn't thought about my career in just this way until Judge Powell put it to me in just these terms, but now that she had, I thought about it over and over again. I had trouble sleeping at night, that's how worked up I was with the decision. Even though the appointment wasn't Judge Powell's to offer or mine for the taking, she put it out like a challenge and I wondered if I was up to it. Charles still remembers spilling out of bed in the middle of the night and finding me in a rocking chair, going back and forth on this thing in a very literal sense. I was wrestling with it.

I ended up applying for the post on the very last day that applications were due. The drill was that each candidate was called in to appear before a panel of Georgia's superior court judges for a roundtable interview. I put in my application on a Friday, and the interview process was scheduled to begin that Saturday. Because I was so late into the mix, they slotted me in at eight o'clock in the morning. Each interview lasted exactly five minutes. They had a timer set up, so that even if you were in the middle of a sentence, in the middle of making a really important point, that was it. It was surreal. In the middle of it all, I remember thinking that it should have been a far more dignified procedure, but this was the way it was.

Long story short: I got the job. They called that night with the news, and I still hadn't told my bosses at Delta. I went in early that Monday morning, at six-thirty or so, to intercept them on their way to the office, and I can say without exaggeration that they were stunned—not that I'd consider leaving Delta Air Lines but that I'd consider leaving Delta Air Lines to become a judge. It didn't make any sense to them. In fact, it didn't make a whole lot of sense to a whole lot of people, and there were still moments when it didn't make sense to me. I still didn't want to be a judge. But a juvenile court judge? Well, that struck me as something completely different. That struck me as the right thing to do, at the right time in my life, for the right reasons. That's a front-row seat to all the ills of our society, at the ground level. That's where you go to make a difference.

Anyway, that's where I was going to make a difference, and I figured the thing to do was take a little field trip to the Fulton County Courthouse to acquaint myself with my new surroundings. I'd been in dozens of courtrooms on a thousand occasions, but I'd never been in a juvenile courtroom, and I needed to see what I'd gotten myself into. And the moment I crossed the threshold into that courtroom I thought to myself, Oh my God! What have I done? It was an incredibly dismal, depressing place. Car thefts, assaults, robberies, murders, deprivation cases . . . and at the heart of each case was a child.

My heart hadn't been broken in quite some time, and here it was about to be shattered in more ways than I could begin to imagine.

## Simple Strategy #4

# Keep Your Word

John Stanford was one of my dearest friends in the world; he taught me a lot about life and living and about leadership and love. One of his favorite expressions was "You've got to love them to lead them," and I used to think it just about the wisest notion I'd ever heard.

John died of leukemia a few years ago after a courageous battle, and lately I can't think of him without thinking also of the lessons in his legacy. He was a retired army general, and following his military career he worked as the Fulton County manager, an appointed position that basically made him the CEO of the entire county; he later became the superintendent of Seattle's public schools. One of the many great things about John was his clarity and purpose, the way he could tackle big problems with small, simple solutions. John had a deep faith in his fellow human beings, in the power of positive deed and thinking, and in his view there was nothing we could not accomplish once we set our minds to it. Commitment, to John, was always key—and follow-through. In the county manager's office, for example, John had draped an enormous banner along the back wall of his reception area with the acronym DWYSYWD emblazoned on it in big red letters. Do What You Say You Will Do. That was John's mantra and his marching order, and he used to love it when people came to his office and asked for an explanation. They'd see these big red letters, DWYSYWD, and have no idea what they meant, and John would take the opportunity to spread his good word.

Do What You Say You Will Do. Nothing less, and nothing more—although in some cases a little bit more couldn't hurt. John used to tell people that the world would be a pretty miraculous place if everybody took those words to heart, and I couldn't help but agree. Think about it: If each and every one of us kept to our word on matters big and small, day after day after day, there'd be no such things as neglect, disappointment, or false hope. Our best intentions would inevitably prevail because there wouldn't be room for laziness, or empty promises, or pie-in-the-sky dreams pinned on little more than fantasy. We'd set our goals and meet them because it was the thing to do—the *right* thing to do, the *only* thing to do.

Do What You Say You Will Do. It's basic, and it takes

me back to the foundation instilled in me when I was a child: Promises are meant to be kept. "Your word is your bond," my father used to say—and of course, in this, he was not alone. The phrase has seeped into our popular culture to the point where it's almost a cliché, but we need to look past the cliché to the meaning behind it. And as parents we need to remind ourselves that we can't expect our children to keep the promises they make to us if we can't keep the promises we make to them. If we say something to our children, they ought to be able to rely on it, don't you think? My parents always kept their word to my brothers and me, without fail. My father in particular was a real stickler about it, to the point where he wouldn't even say he'd make it to a school play or concert unless he was absolutely sure he could attend. "If you say you're going to do it, you're to do it," he used to say. (His words still resonate whenever I make a new commitment.) And indeed, if my father said he'd be someplace, he'd be there; if he said he'd take care of something, he'd take care of it; you could take it to the bank.

Now there's an enormous sense of security that comes with knowing you can count on your parents. No matter what. No matter where. No matter when. It's difficult to articulate what it meant to me as a child, but it meant the world. If my father told me something would happen, then I knew it would be so. I could set my clock by him, and it was a tremendously grounding, reassuring thing. He was very particular about this one piece of parenting, very careful with his words, because he didn't want me or my brothers to ever feel he had let us down or in any way failed to keep his promises, whatever they happened to be. We kids were a priority to him in every respect, but if he couldn't be absolutely certain he'd be able to do something, he'd be real clear about it. He'd tell us he'd do his best but he couldn't promise, and to us kids this was the next best thing. It meant he'd make every effort and then some, and the fabric of our lives could be found in the *and then some*, in the extra efforts he and my mother would make to honor their pledges.

What a valuable gift to pass on to our children. To let them grow up knowing that the adults in their life can be counted on is a blessing that creates endless rewards, and perhaps the most important of these is the responsibility to honor their own promises. We stand and fall on our reputations and, for good or ill, so do our children. Why should your child come to you, or trust in you, after you've repeatedly dropped the ball when he was depending on you to do something? I don't care what it is you've promised, you'd better stick to it. No matter what. No matter where. No matter when. It could be something as simple as showing up for your kid's ball game—but then things pile up at work and you figure you're better off staying late and knocking off a couple of more pages on your brief. In too many households, too many times, the children are made to wait or to understand disappointment. To some parents, each individual disappointment might seem insignificant at the time, but we must realize the cumulative effect. It sets a pattern, and a tone, and reflects a code of behavior that is frankly unacceptable.

The great side benefit to all this is that a child is a whole lot less likely to test a parent he knows he can count on. I'd see evidence of this in my courtroom every day of the week. If a child realized his parents were constantly failing to live up to their word, he might in turn consider it acceptable to break promises of his own—to break the law even.

Let me tell you about Bobby, my car thief. He's in the adult system now, in Georgia, and it tears at my heart the way his life has turned out, but what touches me most about this young man's story is that it turned on a broken promise—or more accurately, on a lifetime of broken promises. No adult had ever kept his or her word to this child, so how could we have expected him to do the right thing on his own? I first caught his file when he was about thirteen years old, and it was one of the thickest files I'd seen. Bobby's thing was stealing cars. He loved cars, knew them inside and out, and he could hot-wire a vehicle so fast he'd be out of the parking lot in the time it took to turn your head. He was small for his age, so he had to scoot down in the front seat to even touch the gas—a mental picture that made the whole business even more distressing.

Bobby came in my first December on the bench. (As you read on, you'll see that most of these "touchstone" cases came to me early on in my career, which I suppose has to do with the learned truth that there are rarely any new and improved ways for kids to get or find themselves in trouble, and after just a short while on this kind of front line I had seen a lot.) Bobby had been in and out of that courtroom so often by that time that all the probation officers and assistant district attorneys knew him. They saw this kid coming and considered going outside to check on their cars. The child had his grandmother there to support him on the day of his hearing, and I took one look at this little kid and his sweet grandmother, and I couldn't see locking him up for Christmas. It may have been what he deserved, but it wasn't what I wanted—and to be honest it wasn't what Bobby needed. What he needed was some straight talk and a plain chance.

Or at least that's what I thought.

I talked to Bobby a bit from the bench, and I talked to his grandmother a little bit, and I've got to admit I was impressed by the boy's charm and good sense. He even had his own kind of integrity, if you can imagine such a thing as a thirteen-year-old car thief with integrity, and I came away thinking that if I could just get this kid to believe I believed in him we could set things right. I laid out a plan that would allow him to go home for Christmas and come back into my courtroom for another hearing just after the new year, at which point we would get to work on some kind of rehabilitation plan.

Juvenile proceedings are closed affairs, but on a delinquent calendar there tend to be a few extra folk milling about the courtroom waiting for their own cases to be heard. Probation officers, caseworkers, assistant district attorneys, public defenders . . . they all kind of huddle in the back of the court with one ear tuned to what's going on at the bench. As soon as I said I was thinking of letting Bobby go home for Christmas, I could see all these heads shaking in the back of the room and feel a wave of disbelief, and I caught bits and pieces of under-the-breath murmurings that all seemed to suggest I didn't know what I was doing.

This may, in fact, have been so, but I was doing it just the same.

The district attorney assigned to the case was a young woman who was herself relatively new to the system. At one point she stood to object to my handling of the matter and said, very dramatically, "Judge, if you let Bobby out, there won't be a car safe in the courthouse parking lot!"

I thanked her for her comments and ratified my thinking that everybody had written this child off, reaching from the back of this courtroom all the way to the front, and I wasn't about to see that mind-set rise to my bench. Maybe there was a reason this child kept driving these stolen cars into all kinds of trouble, and maybe the system had become part of that reason. Maybe this child had started to believe that all of us were expecting the worst from him, so he might as well go out and keep up his end of the deal.

I decided to trust my gut and give this kid a shot.

"Bobby," I said, "I can tell by your file that you've been in a lot of trouble, but this has got to stop."

"Yes, Your Honor," he said.

"I'm serious, now," I said. "It ends today."

"Yes, Your Honor," he said.

"No more hot-wiring cars," I said.

"I promise," he said.

I heard this last and I thought, *There it is*. And there it was: a promise, something I could hold him to, something I could build this second chance around. I heard my father's voice again: *If you say you're going to do it, you're to do it.*

"Listen," I said, "I don't take a promise like that lightly. I'm willing to take a step of faith with you, but your word is your bond." I told him how it was with me and my father, how if you make a promise you've got to keep it, and I suggested we seal it with a handshake. I motioned for Bobby to approach the bench, to make it more personal. It's a gesture I'd never seen another juvenile court judge make, and I'd never made it before myself, but after this one time with Bobby I started bringing my kids up to the bench all the time, often with their parents. If you watch my show with any regularity, you'll see it's an opportunity to really connect with a child or his parents on a specific point. I thought it sent a very powerful signal for these kids to go nose-to-nose with a judge on a kind of equal footing. Like I said, it made it more personal, more human. And Bobby appeared to respond. We shook hands on it. We even did a little high five to seal the deal.

Fast-forward to the first week of the year. The assistant district attorney could see me coming down the hall and couldn't restrain herself from a friendly taunt. "Guess who's back," she said.

I knew right away she was talking about Bobby, but it wasn't what I wanted to hear first thing in the morning.

She followed me back to my chambers as I took off my coat.

"I can't talk about it," she said.

"No," I said, "you can't talk about it. It's an ex parte communication."

"Anyway, you didn't hear it from me," she said.

No, I didn't, I thought to myself. I wouldn't.

Sure enough, Bobby was back on my calendar, this time on arson charges. How he'd gone from stealing cars to committing arson I needed to hear. He'd already entered a guilty plea, so he was in for sentencing, and I cut right to it. "Bobby," I said, "you gave me your word. You promised me if I let you out for Christmas you'd keep out of trouble, come back, get a game plan, start the new year off right. What happened?"

What happened was that he'd set some woman's curtains on fire. He was selling drugs, this woman wouldn't pay, and he set her curtains on fire. "I just wanted to scare her," he said. "I didn't mean to do any serious damage. It's not like I burned down her house."

He copped to the whole thing, and he appeared genuinely ashamed. He hadn't stolen any cars, he'd held up that end, and he ran with the kind of friends in the kind of neighborhood where selling small amounts of drugs to folks in their acquaintance wasn't seen as a big deal to a kid like Bobby, but he hadn't counted on this arson business. He'd meant to keep his word to me, to stay out of big trouble—and yet here he was.

Naturally, I locked him up for a good stretch of time

after this, which at this point was a no-brainer, but what I continued to struggle with was the emotional piece of this boy's story. I couldn't figure out how a well-spoken child could break such a promise—made to a judge, in court, in front of his own grandmother! It was the most bewildering thing.

Over the years, folks have accused me of being all kinds of things as a judge, from a knee-jerk liberal to just a plain old jerk, but one thing I'm not is naive. I may have been a little bit too trusting early on in my career, but I was never naive. I know the score and I know these kids and I thought I could trust Bobby. He was completely honest with me in my courtroom, almost to his disadvantage. He admitted to every single charge against him. We seemed to have made a strong connection. And yet here he was, shredding his word at the first opportunity.

His file told the story. Bobby couldn't keep a promise because he'd never known anybody to keep a promise to him. He was born to a mother who was an alcoholic. When he was an infant, neighbors reported suspicions that they could hear the mother throwing Bobby against the walls. Over time there were additional, substantiated reports of abuse, and Bobby was placed into foster care. An uncle surfaced when Bobby was six years old. He took Bobby into his home and repeatedly sodomized him and otherwise abused him until the child was old enough to articulate what was happening and be returned to foster care. Bobby had been in more than a dozen placements by the time he reached my courtroom—so really, it was no wonder he was in the kinds of trouble he was in.

But still, I wondered. Here I had expected it to resonate for this child that I really cared about him and wanted him to get his life back together, but how do you communicate the importance of keeping a promise to a child who has never known a promise kept? It was a foreign concept to him. At every point in his life, at every turn, he'd known one broken promise after another. As parents, if we expect our children to keep their word, we have to understand what it is that they've been able to rely on. The same goes for judges. I had to realize that this child didn't know what it was to have someone believe in him, to have to hold up his end of a deal. Does this excuse Bobby for stealing all those cars? Absolutely not. For selling drugs? Absolutely not. For setting fire to this woman's curtains? Absolutely not. But why should this child keep his word? No one had ever kept a promise made to him. It's not excusing his behavior, but it's understanding it—and you can't cure a trouble like this until you break it down and find its core. Not to justify the behavior, mind you, but to understand it, to see where it comes from.

Eventually we were able to help Bobby get his life back—for a time, anyway. We found a surrogate big brother for him, a mechanic we thought would stand as a positive role model and mentor and at the same time appeal to Bobby's interest in cars. That worked out well, until the logistics of getting the two of them together proved unworkable, and it was at this point that Bobby's life took another turn. A tragically ironic turn. And while I would never blame this good man who volunteered his time to make himself available to Bobby's rehabilitation, I've never stopped wondering if Bobby looked on the failure of that relationship with the mechanic as yet another broken promise in a long line of adult disappointments. Certainly, after this mentor relationship disappeared from his life, Bobby made another terrible mistake that ended with him being bound over and tried as an adult. He shot a man who was physically abusing his mother—the same woman who used to slam Bobby's little body against the wall when he was an infant. I thought, How ironic. How sad. How unnecessary. It was the last sad stop on a long road of broken promises.

And it got worse for Bobby. His mother's attacker survived the shooting, but Bobby was charged with aggravated assault and tried as an adult, and he came before a district attorney who didn't want to hear about all these broken promises. The judge didn't want to hear it either, and Bobby was sentenced to twenty years. It wasn't justice, not exactly, but there it was, and the reason the district attorney sought such a harsh sentence was that Bobby hadn't shot this guy in the act of abusing his mother. Bobby planned the whole thing, went out and got a gun, arranged the confrontation. It was premeditated, all the way.

It's amazing what children will do for their parents, isn't it? I see it all the time. Parents who abuse their children. Parents who neglect their children. Parents who couldn't keep a promise if their child's life depended on it—which I'm here to tell you it does. Oh, yes, it does.

Sometimes a single broken promise can set a child reeling. Like the eight-year-old boy who stumbled into my courtroom one morning for his deprivation hearing, my first week on the bench. He'd been abandoned by his mother at a homeless shelter several months earlier. Vowing to return, she never came back. Shelter personnel notified the Department of Family and Children's Services, which took temporary custody and placed the child in the agency's emergency shelter, where he'd remained for all this time. Typically, children placed in an emergency shelter are there on a temporary basis until they can be placed into foster care, but in this case the boy had slipped through some of the cracks in our system. It was too long a time for a too little boy to be left in a too busy place like a shelter. Somewhere along the way, this child got it into his head that if he could just hold on until his mother returned he'd somehow be okay, things would somehow turn out all right. His mother had been true to her word before, and he expected her to be true to her word now, and he held on, waiting for the day when she would come back to get him, but she never came back. When the case finally came up and the boy was brought into my court, he must have been told that his mother would be there as well. Clearly he hadn't known about the court date beforehand, but he was old enough to know what going to court meant and that his mother would know what it meant too. He kept looking around, and looking around, craning his little neck to find her. He just stood there alongside his caseworker, looking.

His look was blank, and I strained to read it. Really, I couldn't take my eyes off this child, and as I watched him he started to tremble. Without thinking, I came down from the bench and moved toward the boy, shedding my robe on the way. Like I said, it was my first week on the job—October 1990—and a bailiff approached to turn me back. "Oh, no, judge," he whispered, meaning well, meaning to help. "You can't come down here. You're not supposed to come down here."

I responded with a look that I intended to say, "Get out of my way. This is my courtroom." And I guess that look did the trick, because the bailiff backed away and gave me some room. Realize, at that point I'd been a mother a whole lot longer than I'd been a judge, and my parenting instincts were pretty much overriding any concern for my judicial temperament.

When I reached the child, he was shaking as if he'd been sent out in winter without a coat, so I did what any mother would do. I got down on my knees and I held him. I didn't know what else to do. There'd been nothing in my legal training to prepare me for a moment such as this, nothing in my too brief on-the-job training, nothing in my still limited experience. I was going by my gut, and my gut told me to hug this precious child.

The boy wasn't dressed in his Sunday best or anything like that. He'd been in school, and his caseworker had pulled him out, and he was looking pretty much like any other neglected kid caught in our system. I tried to calm him down as best I could, but he just kept shaking.

Everybody present in my courtroom at that moment had some direct connection to the case at hand. There was the boy's caseworker, the bailiff, the child advocate, the mother's attorney, and an attorney for Family and Children's Services, and all eyes were fixed on this scene. No one moved. No one said a word. No one had ever seen a judge come down from her bench like that.

Finally, I spoke. "I'm not going to lie to you," I said to the boy, who was still trembling. "I don't know where your mother is. She was meant to be here, but she's not here, and I don't know where she is. I wish I did know, but I don't. I will do everything I can to find her, though. I promise."

He wasn't crying, this shell-shocked little boy, and I remember being almost surprised that he wasn't crying. It was a lot for a little kid to have to deal with, but he simply stood there—staring, shaking—his mind probably back at home with his mother. He seemed to calm down after a while, and at some point I realized I had a job to do, so I put my robe back on, returned to the bench, excused the boy and his caseworker, and moved to reset the case. The child would still be snarled in the emergency shelter until a foster home could be located, but there was nothing more I could do for him in court. The file told me nothing about the mother, but I imagined the worst.

Ordinarily it would fall to the caseworker to locate the mother in a case like this and bring her into court, and here I'm guessing that there had been no indication that this woman would fail to appear in court until she in fact failed to appear. I wasn't about to let that happen a second time, or prolong this child's anguish, so I reset the case to my afternoon calendar—the very same day. Here again the bailiff approached me in gentle protest, telling me we can't do this, we can't do that, but again I determined to do things my way. I was too green to know how things usually worked, so I issued a warrant for the mother's arrest, for failure to appear. All these folks were looking at me as if I was crazy—the bailiff, the attorneys—but I was more than crazy. I was furious. I couldn't believe that this woman would let her child down like this, whatever her circumstances, whatever her excuse. I wouldn't have it, not on my watch. And if we couldn't locate this woman and arrest her sorry ass and bring her down in time for my afternoon calendar . . . well, then at least we would have tried.

As it turned out, the woman was fairly easy to locate. She was no longer transient. She now had an address. She had a history of employment. (At one point she'd even been a correctional officer!) And she happened to be at home when the sheriff came calling. I was so new to the system that it amazed me that there hadn't been a more diligent effort before that morning's hearing to ensure that the mother appeared in court, but the court was overwhelmed with so many cases that it was impossible for the caseworker to make an extensive personal effort with each and every one. This child was just one more case in his caseworker's huge caseload, and he didn't surface on his caseworker's radar screen until his case came up that morning.

The mother was brought in for the afternoon hearing, and she was in a stupor. She was wild and belligerent—a crack addict being told what to do by the authorities. I asked the bailiff to get the woman some coffee. That's how green I was, thinking coffee would do a thing to still the rants of a crack addict.

The oath of office I'd taken was still fresh in my mind, and I sat there trying to exercise what I could manage of judicial temperament, all the time struggling to keep calm, to keep focused, to keep this woman from rattling me. In this case too I determined to use Christmas to advantage. It was October, and it seemed to me a reasonable case plan to arrange this woman's treatment, parental visits, counseling sessions, and reviews on the kind of schedule that would allow her to bring her child home for Christmas. Of course, this was a best-case scenario, all the way on the side of hope, but I worked the calendar and set the woman's review for forty-five days, which would have given her a couple of weeks after a thirty-day drug treatment program to get her act together and get back to court and make a case for taking her child home. If she even wanted to take her child home.

Forty-five days later the woman appeared in my courtroom even more belligerent than the first time around. She hadn't been to see her child. She hadn't been near a drug treatment center. She hadn't done a blessed thing. I could order a drug treatment program all day long, but unless this woman was prepared to acknowledge that she had a problem and that she was determined to do something about it, it wouldn't matter. If I sent her under threat of arrest, it would be worthless. Even if she went, she would just have been going through the motions. This woman had two years of college. She'd been a correctional officer. I was amazed at her disdain for the court and her disregard for her child and her circumstances.

Well, I just snapped, and I lit into this woman like she had it coming. I'm not even sure she heard half the things I said, she was in such a fog, but I said them anyway. I was enraged. That the woman had made no effort to see her child, to seek treatment . . . it was unconscionable. The caseworker could only do so much, and back then we didn't have the kind of systems in place, like our excellent Court Appointed Special Advocacy (CASA) program, that would have allowed for closer monitoring of the case between hearings. So I let my anger spill out onto this woman. I told her that one of two things was going to happen: either she'd get herself into a drug treatment program immediately (and that I would drag her down to one myself if I had to), or I would move to terminate her parental rights. The child was eight years old. "I'm not going to let that boy grow up in foster care because you won't make the effort to get treatment," I told her.

"You can't do that," she managed to respond. "He's too old to be adopted."

I couldn't believe that with everything else going on, with everything else I'd put on the table, this was the comment she chose to make, and it set me off. It was as if she'd thought things through in such a warped way that she let herself off the hook, that in her crack-addled head there was no reason for her to be accountable for her actions, because her son was too old and she had no one to be accountable to. I stood, and leaned my hands on my desk, and glowered down at this woman.

"I'm gonna come off this bench and kick your ass!" I said. It wasn't one of the finest moments in the history of judicial temperament, but I was just that angry. "Don't play with me," I said, "because I'll do it."

And I would have (I think) if reason hadn't quickly gotten the better of me. After I'd gotten her attention, and after the shock value of my outburst had subsided, I tried to appeal to this woman as a mother. Clearly I wasn't getting through to her with textbook judicial temperament, so I was up for anything. The woman's name was Kimberly, and I used it here for the first time. I told her that I had two sons, one about the same age as her son. I told her that this was a really painful thing for another mother to see. I told her that if someone came into my house in the middle of the night and tried to take my children, I would fight for them with everything I had, even if it got me killed. I asked her to look at her drug addiction the same way she would a stranger coming into her home in the middle of the night and taking her precious child, because that's what was happening. Her drug addiction had separated her from her child, and she was now at risk of never getting him back.

"It's not a threat," I said, "when I tell you I won't let this child grow up in foster care. I mean it. But it's on you."

All of a sudden this woman saw me as a mother and not just as a judge, and she started crying. It was a real turning point—this incensed judge threatening to come down from the bench and kick her ass turning out to be just another mother trying to do right by her kids. She was hard-core, but now she was crying, and we were able to move forward from there.

The footnote to Kimberly's story was that Christmas didn't happen that year. She went into treatment, relapsed, reentered treatment, found a job, lost it, relapsed again. She got to where she needed to be eventually, but it wasn't a straight line, and along the way there was such a radical change in her appearance that her own son didn't recognize her when she was finally clean and sober. He ultimately went home the following year—on Christmas Eve, no less. I may have missed my target by a year, but at least we hit home, and last I heard the two of them had moved to the Midwest to be with Kimberly's family, and they were finally doing well.

My outburst quickly became the stuff of courthouse legend. I'm told people still talk about the day the rookie judge threatened to come down off the bench and kick some woman's ass, and I'm here to report that the legend is entirely true. You can't make this stuff up—but you *can* learn from it, and the lesson here is that we must sometimes push ourselves to extremes in order to see our way to a solution. My older son, Charles, was about the same age as Kimberly's son at the time, and I kept seeing his face on that little boy, wondering how he might feel if he was abandoned with a lie. True, to be abandoned with a lie and to be simply abandoned add up to pretty much the same thing to a little boy, but it was the lie that got my back up. It was the fact that this tragedy had grown from a single broken promise. There's an unspoken promise between every parent and child, that we will do our best for them at all times. No matter what. No matter where. No matter when.

Now on to a more personal tale, to illustrate how vital it is for us parents to keep our word—not only to our children but also to ourselves. Before Charles was even born, I had an idea in my head of the kind of mother I wanted to be. More to the point, I knew what kind of *working* mother I wanted to be, and throughout my time on the bench I was an advocate for children, working women, and related family court issues. When I began presiding in my television courtroom, these issues remained front and center, especially when I found myself being pulled in all kinds of directions. I had a hectic taping schedule in New York and a full slate of parenting responsibilities down in Atlanta, and there were plenty of times when I felt there wasn't enough of me to go around. It was the classic working mother's lament, and underneath the tug and pull was the promise I made to myself as a young mother, to make my children a priority. To be there for them. No matter what. No matter where. No matter when.

My younger son, Chris, was a starter on his high school football team—and in Georgia, high school football is the kind of big deal that unites some towns and divides others. It was a Friday night. First-round regional play-off game against our top-ranked opponent. Under the lights. Three or four thousand folks with their calendars marked, their hearts set to racing, and their lungs primed for some serious cheering. I wouldn't have missed this game for the world. And then the world came calling. It's funny the way our values are sometimes tested by the strangest temptations, the way push comes to shove to shake us from our plans. In this case the lure came in an unlikely phone call from a New York casting agent, a few days before the game, wondering if I would appear as myself in a music video with Shaggy, the reggae-styled hip-hop rapper who was coming out with a new album. Don't get the wrong idea; I'm not in the habit of taking calls from New York casting agents. This was one of those out-of-the-blue kinds of calls, and it left me thinking, Hmmm. It struck me completely by surprise even as it struck a chord. Why? Well, for one thing, so much of what passes for popular music these days is shot through with hateful, disparaging images, but Shaggy has a much more hopeful message to his music, so the chance to be a part of such a pop cultural positive was enormously appealing. (Don't tell my kids, but I actually listen to all kinds of hip hop music when I am alone in the car!) Plus, for his new single, Shaggy was looking to assemble a group of high-profile, high-energy, high-powered women to appear in the accompanying video, and here they were looking to me. Goodness, I was flattered! And excited! Hey, I'm all for celebrating the accomplishments of each and every successful woman who makes it to the top rungs of her ladder.

As excited as I was, I was careful to put all this into perspective. I knew enough about the entertainment industry to know that at the other end of a long hard day of taping there'd be a split-second cameo that would breeze past so quickly you might have to use a freeze-frame to spot me in the finished video. Or it could be that for all these extra efforts I'd wind up on the cutting room floor. But I didn't care. It sounded like a hoot—and if they wound up using my footage, it'd be great exposure for the show! Of course I would do it. Whatever it took. I'd rearrange my schedule, fly to New York, sit my butt down in makeup, hang with Shaggy, hit my marks, and fly home. Piece of cake.

Trouble was, the video shoot was scheduled for Friday, the day of the football game, and there was no way I was missing that. We were in the play-offs! All during football season, my producers know to book around the team's schedule, which usually means I'm out the door by one-thirty or so on Friday afternoons, in time to catch a flight from New York to Atlanta for the kickoff. On this one point, I won't budge—and why should I? I'll work as hard as they want me to work the rest of the week, go wherever they want me to go, at whatever ungodly hour they want me to be there. I'll take a red-eye or an early bird or a two-seater propeller plane. But if it's a Friday night during high school football season and it's not happening on the fifty-yard line, I won't be able to make it. I'm sorry. I haven't missed a varsity game yet, and I'm not about to start.

Part of the reason I'm so firm on this is that commitment I made to myself back before my kids were even born—to put them first and foremost, whenever practical, whenever impractical. And yet that's only part of it. I could be more flexible, I suppose, if my boys' father—my ex-husband—was going to some of these games, but this part of their lives is not on his radar, which means that if I don't go, they won't have a parent there rooting for them. But that's also just part of the reason. Truth is, I love these games! I've been doing this for a long time at this point, and I'm going to be one sad cookie when high school football is over in our house. Either I'll be out there cheering on someone else's kids or going through withdrawal.

Still, I was torn. I figured there had to be some way to do the video in New York and get back in time for the kickoff—to have my "piece of cake" and eat it too!—and I had all kinds of people making all kinds of calls trying to find a flight that would get me back to Georgia for game time. About the best we could manage was an indirect one-thirty flight to Tallahassee, Florida, which was about a forty-five-minute drive from the stadium, which would have forced me to leave the shoot at noon, but even if I made my connections and the flight was on time, I'd miss the kickoff *and* the bulk of the shooting day. This was a disappointment—a small one, I should add, but a disappointment just the same.

The more I looked at the conflict, the more I realized it was no dilemma at all. I'd go to the football game and hand off the Shaggy gig to another deserving woman, and

I'd probably forget about it until I caught the video on television sometime in the distant future. But other folks weren't so quick to let it go. Most of the people I work with couldn't believe I'd pass up a shot at this Shaggy video. They knew how much Chris's football game meant to me, but they also knew it was just a game. And they also knew Chris. "Ask him what he wants you to do," someone suggested, quite reasonably, except for the fact that this wasn't about what Chris wanted; it was about what *I* wanted; it was about the parameters I'd set for myself and my willingness to stay within those parameters. Of course, if I put it to Chris, he'd have told me to shoot the video. Mom on the fifty-yard line or Mom in a Shaggy video on MTV? He'd have made that trade in a heartbeat, and for him, from his perspective, that would have been the right choice. But not for me.

Chris couldn't believe it when I finally told him about it, the night before the game. I'd been traveling and managed to arrange my schedule to get me home to see that he got his work done and a good night's rest before the big game.

"You'll never guess the call I got this week," I said over dinner, and I proceeded to tell him about my brush with music video stardom.

"They wanted *you?*" he said, emphasis on the sweat-panted, T-shirted, messy-haired woman across the table. He couldn't believe it. "Shaggy wanted you to be in his video?"

I nodded.

My older son, Charles, was equally incredulous when his brother filled him in later that same evening; apparently, turning down opportunities to appear in big-budget music videos with big-name rap artists is not the sort of behavior that passes for normal among teenage boys.

"And you said no?" Chris asked that Thursday night at dinner, making sure he'd heard me right.

I nodded again.

"Why?" he wondered.

"Because we've got a game," I said. "It's the play-offs!"

"But Mom," he said, "we're gonna get our asses kicked!"

He was right about that, but it wasn't the point. We were playing a team from a much bigger school, with a much stronger football program. Our boys would be outmatched and outsized, but to my thinking that was just another good reason why I needed to be on our sidelines, doing my part. Chris might not think he needs me, but the team does. Of this I am certain. I'm one of our loudest and proudest cheerleaders. I've even got my own signature cheer, a simple call-and-response cry that always gets our fans going:

"What do you want?" I'll yell.

"TD!" the crowd will respond.

"What's that?" I'll holler.

"Touchdown!"

It's not original to me, but I've made it my own. (And I've got the strained vocal chords to prove it!)

Chris had a point. Actually, he had a few. He could have called me right after the game and given me the play-by-play. His older brother was coming to the game, so it's not as if he would have been without any family support. A friend could have videotaped the whole game, and I wouldn't have missed a down. Plus, Chris was just a junior, and there was a whole other season of football games for me to look forward to next year. And we *were* going to get our asses kicked.

But none of this, really, was the point. The point was that I would not be shaken from the kind of mother I want to be, the kinds of priorities I want to make and keep, the commitments I have pledged to my family and to myself. Each of us—in every day, in every way—must make an endless string of choices that keep us aligned with our hopes and our best intentions. It's not easy. It's never easy. But the good things come at a price. Sometimes, as was the case here, it's a relatively small price, but quite often the cost is steep. Home. Marriage. Career. Freedom. Life itself. The trick comes in the balance.

Chris was right about one thing, though. We did get our asses kicked. Big time. The final score was 14–45. Note, please, the flip-flop on the custom of reporting sports scores with the winning tally first. I don't buy that, not where one of *my* teams is concerned. What do I care about the other guy's score? At least we scored twice. At least I made it to the game. At least the band showed up and no one was hurt. At least we filled our parents' bus, practiced our cheers, swapped stories on our teenagers for four hours each way, tailgated like there was no tomorrow, and had ourselves a great, good time.

And at least we were all there for our children.

If there's one thing I've learned in all my years on the bench, it's that parents need to be there if they mean to make a difference in the lives of their children. There are all kinds of ways we can screw up once we're there, and all kinds of ways we can flourish, but we must be fully present. It goes beyond keeping our word and stretches to keeping our place. Attendance counts—oh, you better believe it!—and there aren't any truant officers watchdogging us parents to make sure we do our part. It's on us.

All of which takes me back to my opening refrain: Do What You Say You Will Do. If you say you're going to do something, you've got to do it. It's basic. It's simple. And everything good in our relationships with our children will flow from there. Keep your word, and your child will learn to keep his.

Of course, we live in the real world. Stuff happens. Life can sometimes derail us from our best-laid plans, and I'm not suggesting here that we parents must be foolproof in this regard. Not at all. But let's get real: "I forgot" is not a good enough excuse for missing out on a promise. "I didn't get around to it"—also not good enough. Yes, we're human, and yes, we have to say that to our children, but let's do our best to avoid these pat excuses. If there's a traffic jam on the freeway, it's out of our control. If there's an emergency meeting at work, it's out of our control. But we can't simply change our minds if something better comes up. We can't say, "Oh, it's just open school night, it's not that important." We must make every effort, at all times.

And we can't start making little deals in our heads to compensate for the deals we drop with our kids. None of this I'll-make-it-up-to-you-sweetie-we'll-go-out-for-pizza nonsense. You can't wash away your broken promises with treats and kindness. If you tell your daughter that you plan to carve out all of Saturday to take her to the skating rink with her friends, then you've got to do that, even if your golf buddy calls and he has a great tee time and you'd much rather be out on that golf course. You gave your word to your child, and you are meant to keep it.

And God forbid I catch you saying, "I hope you really appreciate this, honey, because I could have been out on the golf course." I hate when parents do that.

The bottom line, though, is the commitment we make to our children to love and support them in all their endeavors; to be there for them at all times and at all costs; to build a foundation for them upon which they can start to build their own lives.

It's all connected, and the thread is your word. Keep it and you're good to go; lose it and you might lose your kids in the process.

# *The Path Taken*

There was a period of time after I first took the bench when I questioned whether I could endure the kind of suffering and tragedy I saw in juvenile court every single day. From time to time I found myself secretly longing for the kind of career where even the tough decisions came easy, where a positive outcome was usually within my control, instead of confronting the endless array of neglect and abuse and despair that found me every day in court. At first the pile of broken lives in Fulton County, Georgia, seemed so deep and the work of getting to the bottom of that pile so overwhelming that I wondered if I could possibly make a difference. And if I couldn't make a difference, then what was the point?

And then, about a month in, I heard a case that turned me around. It was a case with a history. A little boy had been killed by his father some years back. A two-year-old, beaten to death with a stick. The circumstances of the boy's death were not in dispute, but through some clever manipulation of our overcrowded courts and prisons in the adult criminal justice system, the father's attorney had plead his client out on manslaughter charges, said it was just horseplay that ended horribly wrong. The father ended up serving an inappropriately short time in jail, roughly two years. He had an older daughter who had been disabled at birth, and his wife was pregnant with twins at the time of the sentencing, and the man returned to this busy, need-filled household as if nothing had happened.

None of this was at issue in my courtroom on this particular day early on in my career; this was just background. The foreground was this: the wife was now pregnant with the couple's fifth child; the disabled daughter was now six or seven years old and confined to a wheelchair; the twins were now two and being so severely and constantly beaten by their father that the investigating pediatrician was no longer able to document each and every bruise and laceration and fracture. This last was the reason for the family's presence in my courtroom, which happened to be on the twins' birthday. (They were turning three.) It was up to me to figure out what to do about the father, or the children, or the whole lot of them.

Understand, the manslaughter plea came long before my tenure on the bench, but naturally the state brought it up in its case against the father, hoping to demonstrate that this man had a propensity for violence, which seemed clear enough to everyone in that room. Still, I kept looking over at the wife, all through the proceedings, wondering how it was that a woman could stand by her husband in the face of such as this. The man had killed their son. He'd even copped to it, in his own way. How could she continue to live with him? Become pregnant by him, yet again? Support him here in my courtroom? I was new to the job, this was true, but I couldn't imagine that if I worked a lifetime in juvenile court I'd ever see another case quite like this, with defendants quite like these.

The autopsy report from the homicide-turned-manslaughter case was particularly graphic. The child advocate, who was hoping at the very least to have the children removed from the household, became so physically ill at the sight of these pictures that she had to race to the bathroom to throw up. (If I'd thought I could do the same, with anything resembling judicial dignity, I would have been right behind her.) It was horrible, just horrible the way this poor child had been beaten. And yet, through all of this, the wife never wavered. Not once. She stood by her husband and continued to claim he was a good father. "I can't give up on my family," this pregnant woman kept saying, "and my family includes my husband."

There are no juries in Georgia's juvenile court system, which meant that I was judge and jury all rolled into one, and on a case like this it was an enormous burden. There seemed to be no doubt that this man was once again guilty of unconscionable abuse, but the shades of gray came in what to do about it, and what I chose to do about it was pray. I called for a break and retired to my chamber. I'd never done this before, and this one moment of indecision was enough to last me my entire career on the bench. I went immediately to the bathroom mirror, and I looked long and hard at myself, trying to will a solution into my head, and when one didn't come I dropped to my knees and asked for guidance. I didn't know if I could do this. I'd always seen myself as a strong, independent woman. There wasn't anything I couldn't face, but there I was, crying in the face of this. All of this. Every single day. Case after case after case. It wasn't just this one family, I realized. It was everything.

"God," I said, "please give me an extra measure of strength. Please, God, help me to take the energy I give out and move it from grieving for this dead child, and for all these children, to doing what must be done to set things right."

Just then, as soon as I'd given voice to my deepest doubts and fears, I found the strength I needed. Or it found me. It was the most remarkable thing. I've never been one for dramatic, life-altering epiphanies when I come across them in books or movies. They've always struck me as false or contrived. But here it was, a defining moment, make no mistake. I had asked God for an extra measure of strength, and He had delivered. I felt reinforced, rededicated, recharged. I got up off of that bathroom floor and looked at myself again in the mirror, and what I saw this time was resolve. In my eyes, I could see past the grief and focus on the opportunity. It's easy to have your heart ripped out every time you put on that robe, but I couldn't get stuck there. It couldn't end there. I could grieve for that dead child, and there still wouldn't be a thing I could do for him. But out there in that courtroom were twin boys, and for them my response to the matter before me would mean everything. They were suffering. It was their third birthday, and it fell to me to get them safely home.

I still didn't know what I was going to do when I left my chamber, but I returned to the bench with a whole new outlook. A new strength. I continued to search for a solution, but I searched with a purpose. I interviewed the wife at great length. I sat with the children back in chambers and tried to gain their trust and their perspective. At a subsequent hearing, we even had the man's mother flown up from Florida, and she broke down and explained that her son had been severely and repeatedly beaten by his father when he was a small boy.

Now a juvenile court judge in the state of Georgia has a tremendous amount of latitude in the way she hears her cases and the way she disposes of them. The sentencing was often the most difficult piece, because the interests of the state, in meting out a punishment sufficient to deter a repeat offense, didn't always match the best interests of the families involved. This case was a good example. Clearly this man deserved to be put away, although at this point in the proceedings the district attorney had decided not to pursue criminal charges against him. It was also clear to me that this family deserved some kind of chance. And as the wife kept reminding me, this man was a part of this family.

After my initial review of the case, I fully expected to have to remove the children from the home, to tell this woman that if she intended to remain with her husband after he'd served his time on these pending new charges, then I would move to sever all parental rights. I was only one month into this job, and I'd already done just that on other cases, particularly sexual abuse cases where the woman opted to remain with her husband over her children. That would have been the logical move here, the easy move, but now I was vested with all this new strength to try something different, something risky. And here, that bold choice was to root for this troubled family. To give them another shot, the father included. It wasn't up to me to second-guess the justice in the beating death of his older son, I realized, but it did fall squarely on me to put an end to the abuse of his twin boys and to clear a path this troubled man might follow so that he might one day return home and find his family intact.

I ordered that all three children be removed from the home—and that upon the birth of the baby, the fourth child be taken into custody as well. As it happened, the disabled daughter was out of the house for the shortest period—just a couple of months, as I recall, because she didn't make the transition into foster care particularly well and because she was especially dependent on her mother's care. The infant was returned next, after about a month, followed by the twin boys after about a year and a half. All during that time the father was undergoing intensive counseling. The mother too participated in therapy. The children, in foster care, received age-appropriate counseling of their own. In the end the father did serve time, but it was a short stretch. In fact, I think he was allowed to return home sometime before the boys were released from foster care.

It's remarkable the turnaround this man made—and he turned his whole family around in the bargain. He found a way to heal the deeply buried pain in his own life from the abuse he had suffered as a child, then he found a way to heal the pain he'd caused in his own young family. The transformation didn't happen overnight, but it did happen, chiefly because we were able to give it time and room to happen. And I kept close tabs on it. Every holiday while I was still on the bench, the man's mother came to court bringing pictures of all the children and proud progress reports on all of them. The twins were soon enough healthy and happy and making honor roll each year. The father was soon enough back in the home and working at a steady job. My mother kept tabs too. She'd be sure to send over baskets from our church on holidays, and she'd periodically collect clothes and other hand-me-down items and deliver these as well. We were all rooting for this family in what ways we could. I used to keep the pictures of the children in the top drawer of my bench. Occasionally I'd pull them out, especially on days that seemed endless and heartless, and in staring at these smiling, proud faces I'd remind myself what hope looked like. It looked like these good people, reclaiming their lives.

A cynic might wonder why I went to all this trouble. I myself questioned whether I'd done the right thing, then and since. Even with the happy aftermath, I knew that most other judges would have put this man away for a whole lot longer, would have made it all but impossible for this family to ever put the pieces together again. But what I realized that day on the floor of my bathroom was that if I truly believed in rehabilitation, I had to act on those beliefs at every opportunity. If I believed in the abiding goodness of people, I had to do what I could to help them find it for themselves. I had to look, sometimes long and hard, for spirited solutions to dispiriting dilemmas. I had to struggle against the tired, knee-jerk dispositions of too many juvenile and family court judges in too many jurisdictions, and I had to work tirelessly to put together again what had been shattered by neglect and despair. I had to operate on the simple premise that, no matter how bleak a family's prospects, I would approach my work from a place of hopefulness, from an unshakable belief that all could be set right with the appropriate mix of justice and compassion. What I realized, and what has stayed with me, is that the day I couldn't come and sit on that bench with a sense of hopefulness is the day I couldn't do this anymore.

And mercifully, that day has yet to come.

## Simple Strategy #5

# Cheer

Sometimes the potholes in our lives don't appear until they're past the point of filling. So it was in countless families who came before my bench in variously fractured states, and so it is in countless families more who manage to avoid our courtrooms despite the potholes they've neglected for far too long.

I've often wondered what it truly takes to lay a healthy, nurturing foundation for our children, to smooth the paths to purpose and possibility, because if it was such a no-brainer there'd be a whole lot fewer kids in a whole lot less trouble, and what I keep coming back to is the love and support that comes from a caring, constructive family. That's the most vital piece, don't you think? At the bottom of almost every juvenile case I've ever heard, in Fulton

County, Georgia, and on television, you'll often find at least one parent who was so completely uninterested and uninvolved in his or her child's upbringing that it was never any wonder the kid had gone so far wrong.

I meant to be there for my children in every possible way, to be present in every sense, and since both my children were active in competitive sports I never wanted for opportunities to show my support. I had a set of lungs on me that left some folks covering their ears and a set of cheerleading moves that left other folks covering their eyes. I've tried to temper my enthusiasm for my kids' extra efforts, and for their garden-variety efforts as well, but there was a time when I couldn't help myself. I'm sorry, but nothing gets me going like a good ball game when my kids are involved. When Charles was three or four years old, when he first started playing organized sports, I was certifiable. Once, at his very first soccer game, with all these preschoolers bunched around the ball like a swarm, Charles managed to score the very first goal of the game and I just went wild. There was no stopping me. I ran out onto the field, picked up poor Charles, and started twirling him around, and it wasn't until my sixth or seventh twirl that I realized this wasn't so cool. Uh-uh. In fact, this was so not cool I was momentarily embarrassed. (Just *momentarily* embarrassed, mind you, because it took a good few years for me to learn my lesson.)

When the second half rolled around, Charles put in another goal, and this time he flashed me a look that froze me to my side of the field. He was only three or four years old, and he had this killer gaze that said, "You're not gonna come out here again, are you?"

Well, I *was*, but I figured my hugging and twirling would have to wait—this one time, anyway.

You know, it's funny the way time and a little perspective change even something like this. A dozen years later, when Charles started playing high school football, there was a sanctioned time at the end of each game for parents to move out onto the field to mingle with their sons and offer congratulations. If our guys had won, and if Charles had a good game, he used to seek me out and pick me up and twirl me around, same as I used to do to him all those years earlier. In fact, he'd pick me up and twirl me around even if his team had lost. We'd hug and kiss and carry on, win or lose, and in the middle of all that celebrating and whatnot I found time to marvel at the way our roles had reversed and to wonder if Charles had made his own version of the same connection. Indeed, despite the switch, the emotion of these moments was much the same, and it all had to do with the ways a loving family keeps connected, and celebrates each other's accomplishments, and picks one another up—literally *and* figuratively.

Of course, the need to cheer for our children extends far beyond sports and reaches into every aspect of their lives, from school to relationships to small personal achievements like learning to ride a bike or tie a shoelace or play a musical instrument. At all times, at all costs, it's absolutely necessary for our children to hear us cheering them on, to get the message "Hey, I'm here for you. I'm pulling for you. I think you're wonderful." We need to take that message and hang it on the fridge, alongside their school art projects, for all to see.

Most parents take on this role with great good cheer, and they take it on themselves—they don't need some judge to beat them over the head with the obvious. What's not so obvious and what needs hearing is what happens in families where this kind of loving support is nowhere to be found. There's a parade of kids out there, in Fulton County and in the world around, who've never known this kind of care and support and nurturing. All they've known is what they've been told over and over and over—that they're worthless, no good, lazy, shiftless, stupid, not going to amount to anything. Or even worse on some levels, they haven't been told anything; they've just been ignored. I used to sit in that courtroom and cringe at the disconnect between some of the parents and children who came before me, at the complete lack of parental support that I'd see at the heart of most of these cases. These kids had been beaten down for so long, it's almost as if they had no choice but to screw up, just to meet their parents' low expectations, to follow the path that was laid out for them. I kept looking on at these parents, who showed up in court only because they had to, thinking, Why not turn things around? Why not pull your kid aside, even if he's done something wrong, and set him straight? Tell him, "Look, you messed up on this one. I'm disappointed. But I believe in you. You're a good person. You can do better. I'm pulling for you."

*I'm pulling for you.*

It should be on a bumper sticker, because it's one of the strongest messages we can give our children. I tell it to my own kids all the time, and I tell it to the kids in my courtroom at every opportunity, and I mean it. It needs saying. It needs hearing. And—most important!—it needs *doing*.

The disconnect between parent and child reaches beyond our most troubled homes and into the mainstream, and I can almost understand it. I don't accept it, but I see where it comes from. I know what it's like to work (both in and outside the home), to be so completely overwhelmed by the stuff of your days that there's no room in your thinking for how to encourage your children. I've seen thousands of cases involving working poor parents, some of them holding down two or three jobs just to keep the household afloat, and I can certainly see how pulling for their kids didn't become any kind of priority. There was feeding them, and clothing them, and disciplining them, and somewhere way down the list came cheering for them. To some folks it must seem like a kind of luxury, an excess, but if you flip that switch the other way and turn off that support and enthusiasm, then you're bound for all kinds of trouble.

There's a corollary to this parent-child disconnect in the insidious gang structure that permeates our urban neighborhoods—and increasingly our suburban communities as well—and I've got to tell you, it scares the hell out of me. My pastor, Dr. Gerald Durley, put it plainly to me shortly after I took the bench. He had had a conversation with a gang recruiter that troubled him enough to share it, and I pass it on here for the way it darkly echoes my concern. "Glenda," he said, "one of the reasons these gangs are so successful recruiting young people into their ranks is because they can love them better than their families can. They can provide for them."

I thought, What a horrible commentary. And then I thought, Of course! It made perfect sense, in a sickening sort of way. In the tug-of-war between right and wrong, between family and "friends," it was the pull of the street that ultimately prevailed in too many of these cases. It was the lure of these gangs, filling in those spaces where family and community and church might have been.

This gang recruiter was onto something, and I wanted to make a liar out of him. I wanted to send a powerful message, to put my arms around these lost and troubled children, to find ways for kids who couldn't find the love and support and cheer they needed at home to find these things in positive surrogate situations—through Big Brothers/Big Sisters, the YMCA, the church, a mentoring project. In Fulton County, we've got an innovative alternative sentencing initiative in place called Can You Soar?—a twenty-week program funded by my congregation at Providence Missionary Baptist Church that helps build a bridge from the streets back to the community, and it's been far more successful than any of us dared to hope, helping hundreds of kids rediscover the places they left behind. Yes, gangs loom as a new kind of family to these untethered kids, a place where they are cheered for when they're not cheered for at home, and the big trouble is that the cheering is for all the wrong reasons. For all the dangerous, antisocial behavior these gangs encourage, they do appear to foster a disturbingly strong sense of family. It's not the lovingly supportive cheering we long to see in any kind of positive sense, but it's supportive cheering just the same.

Consider the twisted initiation process in most urban gangs. In many cases a new recruit will be sent off to complete his or her "pieces of work"—that's how it's known in gang parlance. So, for example, one piece of work might be to carjack a car—or for girls, the piece of work might be to sleep with so many gang members, have oral sex with so many gang members, and so on. It's all part of the long process of gaining trust and fostering loyalty and becoming accepted into this surrogate family. When your piece of work is completed, you are praised for that piece of work, and it's criminal and warped and dangerous and yet that's how it goes, and it plays on the need of these troubled kids to be acknowledged and applauded. They are, at last, accepted. They belong. Folks go out of their way to sing their praises.

One of the most disturbing situations I've seen in court involved a girl who had been in a gang but hooked up with a new boyfriend from a rival gang and wanted out. Now it's not as if you're in the Girl Scouts and you decide you don't want to do scouting anymore. You don't just walk away, especially not to a rival gang. And in this girl's case, she had been a very plugged-in member of this particular gang. She ran with some of the key people, was fully initiated, knew all kinds of details concerning gang business and relationships. She'd become so fully immersed in the upper workings of the gang that her decision to leave put her at risk. Typically, with girls, your standing in the gang hierarchy has to do with your boyfriend's standing, and this girl had reached the top in such a way that there was no way out.

So they went looking for her. First place they looked, naturally, was at her mother's house, and when she wasn't there they beat the daylights out of the mother, but the mother hadn't seen the girl, didn't have any idea where she was or how to contact her. Of course, that's a whole other issue, how a mother of a teenage girl can go for weeks without knowing her child's whereabouts, but that's often how it goes. Anyway, the woman picked herself up and went down to the courthouse looking for her daughter's probation officer. Naturally the girl had been in a mountain of trouble already, so she was already in the system, and her mother had folks she could talk to, to lay out this new situation. Her thought—the same as the probation officer's, the same as mine—was that if these gang members found her daughter they were going to kill her, and fortunately the police were able to go out and find the child and bring her back to my courtroom. The probation officer and the public defender were asking me to take the girl into protective custody because if I let her back out on the street she'd be as good as dead. The mother told me about her beating, which was a classic "message beating"—it wasn't savage, but it was brutal enough to get the point across and to scare the mother into seeking help.

I felt I had no choice but to take this girl into protective custody, and yet I couldn't leave her locked up indefinitely. Understand, it's not as if this child had seen any kind of light and turned her righteous back on gang activity. She just wanted out of one particular gang so she could run with another. So I wasn't about to cut her any slack for turning over a new leaf. Still, I didn't want to see her killed, or otherwise threatened, or to cut off her chances of ever turning things around, and there was a whole series of things we had to weave together to get her committed to the state, to keep her away from people who knew of her gang affiliation. In fact, after she was in jail for a couple of months, a couple of girls called her by her gang name, which of course was a not-so-subtle signal: *You're not fooling anybody, we know who you are*. So we had to move her to another facility.

It was just a horrible, horrible situation all around, and the point of the whole mess in this context was the extent to which this young girl had become enmeshed in the gang. She chronicled the whole sordid ordeal of her initiation, which included being sodomized by several members of the gang. The most chilling piece of work was her final assignment, when she was dispatched to kill someone—her own mother! It was at this point that the girl finally wanted out, and what I kept hearing between the lines of her account was the sad, scared voice of a sad, scared child, so desperate for acceptance and attention that she grabbed at it in what ways she could. This girl eventually pulled away, but a lot of kids don't.

The gang culture permeates every ethnicity and social and economic class. It's all around, and if we are not out cheering for our children, in every way, at every opportunity, there'll be all kinds of negative influences to take our place. Consider another disturbing story: A local senator is all set to write a letter of recommendation for a high-achieving high school junior's commission to the Air Force Academy. By all outward appearances, this was a great kid: ROTC, model student, written up in the paper for various good deeds and high marks. Every parent's dream, right? Well, the summer before his senior year in high school, this young man just disappeared. Fell off the map, as far as his mother was concerned, and after a couple of weeks, he was picked up on a major drug trafficking charge. The scam, as it was laid out in my courtroom, was for this young man and his partners to deliver a considerable cache of drugs to a location in Tennessee and return to Georgia in a fleet of luxury cars. It was a classic drug-money laundering scheme dressed up with these high-end vehicles, and this otherwise good kid, with no prior record and all kinds of respectable opportunities in front of him, managed to get himself caught.

He appeared before me as a very mannered, very articulate defendant, but he wasn't very forthcoming. He was caught red-handed, across state lines, and there really wasn't anything he could offer in his own defense. His attorney didn't ask for a trial, and I got the whole story at the sentencing hearing. I had a very thin file in front of me, and I remember thinking how unusual it was to see a kid with this kind of background on major trafficking charges; really, you don't usually see a first offense on something that serious. His attorney played it the way I thought he would, asking for leniency in my sentencing, pointing to the young man's outstanding record, calling this a very bad indiscretion that didn't have to cost his client the rest of his life. The boy's mother was weeping, just frightened out of her mind over what her child was facing. The boy's father had come up from Alabama, and he was fairly pragmatic about the situation—upset and disappointed but nevertheless realistic.

On first blush, I was inclined to agree with the lawyer. This was a very bad indiscretion—although I might have tossed a few more *very*s into the mix. I thought, God, this is awfully serious, and yet I had to think there was some sort of way to salvage this young man's situation, to build on the fine foundation he had laid before this recent turn. Clearly there was a question of whether he'd be admitted to the Air Force Academy after something like this, no matter what my sentence was, but I wanted to give him a shot at a college education, at turning his life back to its previous course. Truth was, if I shipped him off to the state, he'd never get his life back. If I came up with a tight probation scenario and a tight review schedule, perhaps he could return to form. It would be on him, but that path would be open. Plus, he'd already spent some time in jail before he turned up in my courtroom, and I thought that short stint might be enough to teach him a lesson and set right things right.

I talked to the parents a little bit to see what kind of support there'd be at home to help with the turnaround. The father was willing to take the boy back to Alabama with him, if it was determined that this would be a good thing. The young man wanted no part of his offer, but at the same time he refused to go back home with his mother, and this is where the case took a stunning turn. All along I'd been thinking the child would welcome the chance to go home with his mother, especially against the alternative, which would have been a long stretch in a state institution, but when I put it to him straight he was dead against it. "She's not my family anymore," he said, cold, and his words just about took the wind out of everyone in that courtroom. I wasn't even the child's mother, and it felt to me as if someone had kicked me in the chest. I looked across the courtroom to his mother, and she was devastated—crying, nearly wailing. It was almost as if she was being forced to watch her son being abducted.

I put it to the boy again, just to make sure we were clear, and he put it right back. "She's not my family anymore," he repeated, almost robotic, as if the line and its emotionless delivery had been rehearsed—which I have no doubt it was.

This was the first any of us on the side of the law knew that this young man had been recruited by one of the local gangs. Up to this point it had been generally assumed that he had simply stumbled onto this scheme and saw it as a way to make some quick cash, that it ran no deeper than a bad decision. The young man's attorney had offered no indication of any gang involvement, presumably because he had no indication himself. But here it was, plain as day. The kid was running drugs for the gang, and the gang was his new family now. It was almost as if he'd joined a cult, the way he suddenly appeared before us, helpless to see his situation clearly out of a perverse loyalty to the gang.

I wondered how they'd gotten to this kid in just a summer and so penetrated his soul. I wondered what pieces might have been missing from his own puzzle and how these gang leaders knew when and where and how to fill those resultingly empty spaces. Realize, this wasn't some struggling, unfocused high school student. This was a disciplined kid, a good student, a model citizen, excelling at the highest levels. (You're not striving for the Air Force Academy if you're not willing to work hard.) I could see everyone in the courtroom shaking their heads—the public defender, the probation officer, the bailiff. . . . These were folks who had seen a whole lot of stuff, but they hadn't seen this, and none of us could think what to make of it. It shook us, I think, because it hit so close to home. Somewhere in the back of our thinking we were all wondering what might keep our own good kids from this same bad circumstance.

I chose to take a hard line. Or to put a finer point on it, I had no choice but to do so. I said, "All right, if that's the way you want to play it, we'll reset the matter for another thirty days and lock you up until then. Let's see how you feel after that."

"Lock me up if you want to, Judge," he said, matter-of-factly, "as long as you want to, but eventually you'll have to let me out, and I'll tell you what, I'm never going back to my mother."

There was nothing in the young man's family history to suggest this kind of break. He'd never been abused, never been neglected. Ultimately, though, he didn't have the support he must have needed. His dad was out of state and only superficially involved in his life. His mother, who had a good soul and a good heart, was a hardworking, persevering woman, struggling just to keep things going but managing to keep and maintain a loving, comfortable home. From her perspective, she had done everything right—and she had!—and now she had this high-achieving child who was getting all kinds of positive attention and feedback in his community, and possibly heading to the Air Force Academy, who clearly felt there was something missing in his life, something he wasn't getting at home. And whatever those things were, he seemed to have found them with this drug-running gang, to such an extent that he was prepared to walk away from the rest of his life, from that all-important path to purpose and possibility, and head down this uncertain road to no good.

Looking back, I realize it makes sense for gang leaders to target young men like this defendant. They don't need dummies helping them to run their multimillion-dollar drug operations. They don't want kids who can't add, kids who aren't savvy or slick, kids who don't know the score. A high-achieving, well-spoken kid like this Air Force Academy candidate was just who these guys were looking for, and that's one of the most distressing aspects of this case. Maybe there was no way to avoid it. Maybe there was nothing this young man's mother could have done to steel her son against such advances, against the get-rich-quick glamour of the gang lifestyle, against the abundance of love and support from this new community, against the promise of easy answers and no consequences. These recruiters can fill up the potholes in a young life like nobody's business, and that should put *all* of us on the alert—from the barely-going-through-the-motions parents of troubled kids in broken, dysfunctional homes straight on up to the over-the-top involved parents of striving, achieving kids who appear headed for great things. And the burden's not just on parents; it's on all of us who love and care about children and the future of our civilized society.

I ended up committing this child to the state, and it just about broke my heart, but I felt I had no other choice. He made it very clear that he wasn't going home, and that if I released him after a period of detention or rehabilitation, he wasn't planning to stay there. He wasn't going to live with his father either, even though he expressed none of the robotic coldness to his father that he displayed toward his mother. The state was a last resort I never wanted to pursue for the way it took matters out of my hands. If I found a way to detain a child within my jurisdiction, I was able to be involved in treatment; I could order certain things to happen; I could monitor a child's progress in the kind of hands-on way that frankly never happens at the state level. But there was no cause to keep this young man on such a close leash if he had no intention of working toward a positive goal. Our detention facility was designed only for short-term stays, and I couldn't see putting him back out on the street and reuniting him with his new gang family.

These "pieces of work" are a killer. Some of these lost souls get so blinded and intimidated by the gang-initiation process, they ignore some of their hardwired concepts of good and evil and right and wrong just to gain acceptance. It's one thing for a kid to attempt to justify breaking the law in what he might perceive as a "victimless" crime, such as a school yard drug transaction or an orchestrated shoplifting, but these kids were often dispatched with weapons and marching orders to kill.

Now I realize that these examples are probably a little extreme for most of my readers, for whom this kind of elemental disconnect may not resonate. But sometimes our best lessons are learned from the extremes. Positive reinforcement encourages positive behavior, and there can never be enough of the stuff to go around. Put a kid like this in a loving, nurturing household and his life plays out another way. Or at least you up the odds that his life plays out another way, because after all, when you break it down, it's all about the odds, yes? This parenting business is all about increasing your child's chances to find his or her own positive path, and on this score I maintain that the single best way for parents to tilt these long odds in their favor is to celebrate their children. Sing their praises at every opportunity, and even when you can't find something to sing about you can at least hum a few bars. In our house, what this means is celebrating everything, to where we sometimes go out of our way to find a reason for the applause. Even no reason at all is reason enough. And as I wrote earlier, the cheering doesn't stop when we leave the ball field. There are reasons to cheer everywhere we look—from a positive interaction with a clerk in a store to the responsibility to pick up a room before being told.

Over the years, I've found that one of the best places for a lot of this full-throttled cheering is the family car. Think about it: There's no better time to capture your kids' attention, or for them to grab yours, than when you're all safely seat-belted in the car. There's no place else to go—and frequently nothing else to do. As my kids moved up through middle school and high school, our household became a very busy place. You know how it is. Between homework and chores and extracurricular activities, there never seemed to be enough time to sit still with one or both of my boys long enough to run through what some folks might see as the more mundane aspects of their days. So I hit on the notion of designating our time in the car for this all-important business. I used to drive them to and from school, and we had a rule that in one direction they couldn't have the radio on. Naturally they didn't like this rule much when I first put it in place, but I gave them the choice, morning or afternoon, and beyond that I told them they'd have to abide by my decision. Eventually the griping died down, and our trips to school were soon filled with a running commentary on this and that. Together we created a time and place where we could fill one another in and egg one another on. There were opportunities to second-guess, and to troubleshoot, and to congratulate, to the point where some days we'd keep the radio off in both directions.

I can't stress how important it's been in our family to squeeze out these little moments of time, these points of connection. The other great locus has been bedtime, and even though my kids might cringe over such a public admission, I'm here to report that I still tuck them in at night. Chris, anyway. Charles is off at college, but when he's home, he gets the same treatment as his younger brother. And do you know what? Neither one of them seems to mind. In fact, I think they like it. Their friends ride them about it whenever it comes up, but they shrug it off. My own friends ride me about it too. They tease that I'll be at a loss when my kids are grown and married, but I'm fully prepared to tuck in the whole lot of them. It's just like that key time in the car. It's quiet. There's nothing else going on. It's the end of the day—the perfect time to report and reassess. Really, it's been a wonderful barometer in my relationships with each of my children, and I won't give it up without a fight. And for the next while at least, I won't have to.

Lately, with Charles out of the house for the most part,

Chris and I don't save our serious talks for bedtime the way we used to. Now they come up when they come up, and he doesn't have to worry about his brother listening in to anything that he wants to keep private. But I still tuck him in because . . . hey, you never know, right? Once, after I'd been away from home for a long stretch, I came home late at night and woke him up. I'd been gone an extra couple of days, and my plane was late, and he'd meant to wait up, but it was about midnight or so when I pulled into the drive. There'd been a lot going on with him at school and with a few of his friends, and I knew he was sitting on a whole bunch of stuff that he needed to share, so I shook him awake. I figured if he wasn't up for it, he'd turn right over and go back to sleep and we'd catch up at breakfast, but he sat straight up and started gabbing. Next thing we knew it was two o'clock in the morning, but he needed to talk and I needed to hear it. The next day was a school day, but sometimes you have to roll with a moment like this one. The central issue of this late-night tuck-in will have to remain between Chris and me—because after all, the private nature of these talks is what makes them special and productive. (Goodness, what kind of mother would I be to betray his confidence in such a public forum?) The important point is just that something was troubling Chris and he needed a sounding board. He couldn't understand a certain type of behavior demonstrated by some of his friends—he was furious about it, actually—and at first I silently congratulated him on his well-placed sense of moral indignation, but then as we moved away from this dilemma and onto the next one I realized that silent congratulation doesn't always do it. Here my youngest son, all of fifteen years old, was showing the kind of maturity and decency I'd wished for him since he was an infant, and he needed to hear how proud I was of how he'd turned out, how he was turning still. And he needed to hear it loud and clear.

So I cheered. Granted, it was a two-in-the-morning sort of cheer, something completely different from the hooting and hollering I've been known to register on the ball field, but there was no mistaking it for anything else. I turned the conversation back onto that other boy and the younger girl and commended Chris on his compassion and insight. There was no chance of waking the neighbors, but I couldn't have given him the message more forcefully with a bullhorn and a marching band. It may not have been the most full-throated cheer in history, but it was certainly full-hearted, and in the quiet I reminded myself that sometimes the most meaningful praise is what doesn't need saying. Sometimes a look or a touch are all you need to send the message home.

## *Cases in Point*

It can happen in an instant, even to the "best" kids.

Somebody pulls up after school driving a brand-new Lexus and offers your child a ride. Your son—let's call him Billy—doesn't want to be seen as uncool, so he considers it. He knows you've told him to take the bus, but he runs the situation in his head. *Who's gonna know?* he thinks. *What's the harm?* He figures he can cruise around with his buddies, go to the mall, buy a few burgers, holler out the window at girls, and still be home before the bus would have reached his stop. And even if he's a little late, he'll still be home before you get back from work. Just to make sure, he'll have the guys drop him off a couple of blocks away and walk the rest of the way in case you've gotten home early. He's got it all figured out. He'll get his joyride and you'll never know he went against your wishes.

End of story? Sometimes. But sometimes—too often, in fact—it's just the beginning of a sad, unfortunate ordeal. Sure, the scene can play out pretty much the way Billy imagined it, but it can also play out in any number of ways. Regrettable ways. Life-changing ways. Life-*ending* ways. I've seen it all and then some. What if the car Billy's friends are driving turns out to be stolen? What if there are drugs in the car, and the driver runs a stop sign, and when he's pulled over the policeman gets a whiff of marijuana and now has probable cause to conduct a search? What if there's a gun on the floor beneath the front seat and the butt has been somehow jostled into view? The point is that good kids can frequently find themselves in bad situations because they feel the pressure to be like everybody else, to go along with their peers even if they're headed down the wrong road.

The consequences can be devastating. What happens if Billy decides to do the driving? None of his friends have their licenses, but they've managed to get behind the wheel—and if the car's stolen, or "pinched" from someone's unwitting parent, these kids aren't caring all that much about who's doing the driving. Let's play this out to its worst-case ending. Billy doesn't want to be seen as any kind of loser who's too chicken to drive, so he takes the wheel and ends up wrapping the car around a telephone pole. Two people die. All of a sudden, spinning off of this one bad decision, Billy's facing charges of vehicular homicide. He's a good kid. He's never been in trouble. He's a track star at school, a straight A student. But he's in big trouble now.

And what if it's your car Billy's managed to pinch? Then, on top of these horrible criminal charges, you'll also be dragged in to face some pretty severe civil charges, from which you might never recover. You could end up working the rest of your life to deal with a judgment that far exceeds any possible insurance policy. Ask ten adults with high school kids of driving age what kind of automobile insurance they carry to cover just such an eventuality, and there's a good chance only two or three of them could tell you with any certainty, and among those few folks it's unlikely that even one of them has sufficient coverage. People just don't think along these lines—and they should. It's situations like these that lead to parents pressing charges against their own children for stealing the family car. Okay, I understand that as a simple, practical matter, it makes more sense for a child to have his day in juvenile court than for a parent to be exposed to a multimillion-dollar civil claim—and yet I can't imagine the emotional scars some families carry after having been split apart by this kind of scene.

Every week, or just about, I'd hear a joyriding case in Fulton County and think to myself, That could be one of my children. As often as not, these were good kids, with involved parents, their young lives turned on the back of one bad decision. Just one. Or maybe they'd made a series of bad decisions and had yet to be burned by a single one of them, but the heat from this one was going to scorch. Oh, you better believe it.

A mother leaves her car at the shop to be repaired and rents an awesome sports car so she can get around. She's thinking that for a few dollars more she can drive around in style, at least for one day. Her two teenage boys are thrilled with the sports car. That night, one wakes up the other and convinces him to take the rented car out for a spin. *Who's gonna know?* they ask each other. (After everything I've seen and heard in my time on the bench, I can't imagine a more common phrase: Who's gonna know?) Sure enough, they manage to roll the thing out of the drive without waking their parents, and sure enough the driver loses control of the car and his brother is killed in the resulting crash. A parent's worst nightmare, right? One son is dead and the other is facing vehicular homicide charges—and a lifetime of knowing that his brother died at his own hand.

A young girl gets into her boyfriend's car. They don't have any special plans. They're just hanging out, driving. At one point the boyfriend pulls into a convenience store parking lot, tells the girl he'll just be a second. He needs to get a soda, maybe a snack—does she want anything? The girl remains in the car, the motor running. A couple of beats later, the boyfriend comes racing from the store. He jumps in the car and pulls away. Next thing the girl knows, there's a police car chasing them. There's a real, high-throttled, high-speed chase, just like in the movies. Eventually the boyfriend pulls over. Turns out the boy had meant to hold up the convenience store, the robbery went bad, and he ended up shooting the clerk behind the counter. The clerk subsequently dies. The girlfriend had no idea, but she's up against it just the same. She's arrested and charged as an accomplice in a felony murder. She ends up being convicted and sentenced to twenty years in prison. Understand, the law in Georgia doesn't distinguish between the kid who pulls the trigger and the kid who stands as lookout; they're both guilty of the same crime. And yet this girl claimed she had no idea, and all accounts indicate she was probably telling the truth. This wasn't my case, so I don't mean to second-guess the presiding judge, but this struck me at the time as an awfully harsh sentence. Twenty years! She'll be almost forty years old by the time she gets out. This girl had no prior record. She was by all accounts a good kid, but I suppose it's possible that the judge didn't buy her story. Maybe something didn't wash. Or maybe the judge wanted to send a powerful message.

One story bleeds into another, but at the bottom of each there's the same lesson. We can't afford to let our children make these bad decisions. We can't afford to let them hang out with bad-ass kids. We can't let them cave in to whatever social pressures they're facing at school. Now I was a teenager once myself. (And not that long ago, thank you very much.) I got into my share of trouble. But as parents we must remind our children at every turn that what seems to them to be harmless fun can go horribly, devastatingly wrong. No, we can't choose our children's friends. No, we can't watchdog them twenty-four hours a day, seven days a week. No, we can't stand by their sides and ease them past every social dilemma that finds them at school or out on the town with their friends. But we can drive home for them the simple, alarming truth that bad things happen to good kids. They do, they just do. These kids can't expect to keep making bad choices and not be tainted by them.

All of which takes us back to Billy. There are so many ways that one joyride can play out, leaving poor Billy wishing he'd listened to his parents and taken the bus. One of the outcomes I heard most often was the driver getting pulled over for a minor traffic offense and the cop noticing drugs on the front seat or dash. In such a scenario, in most states, the law suggests that possession only applies to those persons within reach of the illegal items, but not all police officers take the time to apply this standard. The simplest move is to arrest everybody and figure everything else out later, right? Sometimes the driver will cop to the possession, and in these cases the district attorney will most likely dismiss the charges against the passengers in the car, but what happens more often is that the busted friends start pointing fingers at one another, nobody wants to take the full force of the heat, and everybody gets dragged down in it. Our Billy might have had no idea there were drugs in the car. It's possible these kids weren't even stoned and that the driver hadn't been driving recklessly. Maybe he was pulled over for a broken taillight. But Billy is busted just the same.

Talk about being in the wrong place at the wrong time—and yet it's at crossroads like these that we find our best opportunities to communicate with our children. Goodness, we can talk to our kids until we're blue in the face and fresh out of gas, and we still won't make our point. But when a story like this hits home in your community, among your child's circle of friends, you should be all over it. Seek these kinds of cases out in your local newspaper and hope to God your child has the sense to learn from the mistakes of his peers. And hope to God again that the next story won't have his name on it.

## Simple Strategy #6

# Make Money Matter

Money matters. Oh, yes, it most certainly does. To our children especially, it matters in the obvious ways, and it matters in some subtle ways, and it matters in ways we can't begin to see for trying.

Parents who think their children have plenty of time to learn about managing their own money and establishing money-based priorities are in for a rude surprise, because it's an issue by elementary school. Whether we like it or not, whether we guard against it or not, even our youngest kids begin to see the world in terms of dollars and cents, and the world looks back with its own kind of price tag, and what happens in the back-and-forth is a whole mess of complicated emotions and difficult choices and mixed signals. It's no wonder our kids sometimes get the wrong ideas about what money means and how it matters—or worse, no idea at all.

Get to the bottom of almost every juvenile court case—involving drugs, gang violence, truancy, deprivation, domestic violence—and you'll find money somewhere at or near the bottom line. Therefore, downloading onto our children a clear, value-laden understanding of, familiarity with, and appreciation and respect for money is all-important. Goodness, it's vital, and it's never too soon to start.

In our household, what this meant was a hands-on approach. Very early on, just to offer a basic example, I put my boys on a clothing budget, which I thought would help them to prioritize their "wish-lists" and learn the value of a dollar besides. Frankly, I also thought it would get them off my back. I don't know how it is in your house, but in mine there was an endless wail of "I want," "I want," "I want." Every time we went into a store, or passed a billboard, or sat through a commercial on television, there was something my kids desperately needed to own—and right away! They had no concept of how much things cost, or how much money we had to spend, or how to weigh the merits of one impulse against another, so around the time my older son, Charles, was eleven or twelve, I hit on this notion of a clothing budget. Actually, it was more of an allowance than a budget, because I thought it was important to physically hand over the money and let them manage it—to the extent that they could, of course. Twice each year, once in the spring and once again in the fall, I'd give each of my kids what I thought they needed. I bought all the essentials, like school uniforms, pajamas, underwear, church clothes, and overcoats, and they had to buy everything else. If there was a cool new jersey they wanted, it came out of their budget. A stylin' new jacket? Their call entirely. Jeans, sneakers, sweats, ball caps . . . it was all on them, and it forced them to think things through. And it had the added benefit of liberating me from the ranks of the wardrobe police. I'd no longer have to look at them as if they were plain crazy when they came bugging me for a new pair of Nikes just a week or so after I'd bought them a perfectly good (and perfectly expensive) new pair. It would be on them.

Their allowances grew with them. In elementary school I earmarked three hundred dollars for each of the two shopping seasons; in middle school that was ratcheted up to five hundred dollars; and in high school I gave them each eight hundred dollars, twice a year. It sounds like a lot of money, but I figured it out once and I managed to come out ahead on the deal. Do your own math. Sneakers, warm-ups, visors, accessories . . . it adds up, and it didn't take my kids long to realize that they could get along just fine without some of the things they thought they desperately needed to have. It forced them to make some choices.

Let me tell you, it was one of the smartest moves I ever made as a parent, and a couple of life lessons came with the bargain. One of the toughest for my kids to learn was perhaps the most essential: *When it's gone, it's gone*. If Chris blew what was left of his allowance on a ridiculously overpriced basketball jersey and a new one came out just a few weeks later, he was plain out of luck. He had to wait until his next allowance installment. That's the way the world works, and if the rest of us working stiffs need to get this message, then our kids need to hear it too.

Another lesson: *Let the buyer beware*. How many times have you heard parents moaning about this or that toy that somehow manages to look good on a television commercial and yet comes out of the box from the store looking like a cheap imitation? Well, we grown-ups don't get our money back when we've been suckered, and our kids need to understand this. Sometimes, if it looks too good to be true, it may well be. Charles went off one weekend with family friends to a flea market and had himself a shopping spree with all the knockoff designer clothing they had laid out on the tables there. I wasn't with him at the time, to offer up my two cents, and he just went wild. He bought a whole bunch of stuff, at a fraction of what the real deal would have cost him in the stores, but one run through the wash and everything was headed for the Goodwill box. The stuff was so poorly made it was all but worthless, which meant that even at these low prices Charles had overpaid. As a mother, it was difficult to stand by and watch, because he'd run through most of his allowance for that season, but a lesson earned is a lesson learned.

I'll let you readers in on a little secret: A few weeks after the flea market fiasco I went out and bought Charles some nice clothes, as a little pick-me-up present. I wasn't about to buy him out of his mistake and replace those items one by one, but I figured it wouldn't hurt just to see him smile. And underneath this admission is another important lesson, reminding us parents that it's sometimes okay to soften our hard line. Yes, we must say what we mean and mean what we say, but we must never lose sight of the fact that our children hurt easily, and we need to allow ourselves some wiggle room to make a few small repairs.

There is no such thing as easy money, and our kids need to get this message as quickly and as fundamentally (and as forcefully) as possible. The ability to earn, control, and value money is an essential tool, and this was never more apparent than in the cases I heard in juvenile court. Everything, it seemed, turned on money. Drug dealing, shoplifting, prostitution, gang violence . . . money is indeed at the root of our street-level evils. Once again, the cliché rings true, and it rang with particular clarity when I heard a deprivation case involving a sixteen-year-old girl. This was a child who could have come before me on either side of the court—on the delinquent side, for some of her suspect activities that at this point had amounted only to petty infractions in her file, and on the deprivation side, for the complete lack of parental supervision in the girl's home and for the alarming delinquency record she was building as a result. She came before me on the deprivation side, on a review. There was no open delinquent case against her at the time, but there might have been. Indeed, there should have been.

We got to talking. This child looked much older than

her sixteen years, and she was hardened in ways I found deeply troubling. She told me she used to sell drugs, and she very proudly laid out her code: She only sold drugs on the weekends and only to the wealthy kids on the North Side. She laid this out as if her distinctions mattered—which to her, I guess they did. According to this girl, she was able to clear one thousand dollars, cash, each day she worked (Friday night, Saturday, and Sunday), which on an annualized basis amounted to a whole lot more than I was earning at that time as a juvenile court judge. I thought, What do you say to a kid like this? How do you get her excited about flipping burgers or bagging groceries for minimum wage when she's looking at this kind of money? How can you get her jazzed about going to college if all she could hope for at the other end was a career that would bring home less than her casual dealing? The more I thought about it, the more I wondered what there was to say—and yet here she was, on the other side of her bad decision.

Even though she gave up dealing drugs on her own, after being forced to watch the execution-style murder of her boyfriend in a drug deal gone horribly wrong, it didn't change the central dilemma this child presented to me that day in court. How do you counter the false promise of easy money to kids on the street? How do you instill any kind of work ethic when no ethics at all are required to earn piles and piles of money? There are very few arguments against making one thousand dollars a day that a kid is willing to hear. Of course, a thousand dollars is a little on the high side, but even a hundred bucks a week is hard to argue with, especially when you account for taxes and regular hours and all that good stuff. Even "good" kids have a difficult time with this equation, because they know the score. They see that the odds are weighted all the way in their favor. They know they'll most likely get away with a twenty-dollar marijuana exchange because the cops are too busy chasing down the cats on the hundred-thousand-dollar cocaine deals. They know how time-consuming it is for a police officer to make a bust stick in court, and they know that even if they do get caught and prosecuted, the system is so overwhelmed that they'll spend very little time in jail, if any. They like their odds a whole lot better than they like the idea of punching a clock down at Old Navy, stocking shelves.

Down the line, the arguments in favor of signing on for a traditional, law-abiding minimum-wage job match up with perfectly sound arguments against it, and we're left with a traditional risk-reward scenario without a moral compass. Sad to say, there's no reaching some of these kids on the moral or ethical side of this conversation, but that doesn't mean we stop trying, and about the only success I've had from the bench is to turn it into a conversation about race and opportunity. I tell these kids, busted on the bad ends of their bad decisions, to take a long hard look at what their actions are doing to the African-American community, if it so happened that these kids were African-American. Money is almost always the through-line to these missteps, and I try to get these kids to understand that. They don't think in terms of right and wrong. They don't see any kind of big picture. They only care about what's in it for them, which usually translates into how much money they can make on the deal. They look at me as if I'm from some other planet, but I press on. I ask them what they would do if a flatbed truck came rolling into their neighborhood in the middle of the night, with a bunch of men covered up in white sheets on board, opening fire on everybody they saw. Women and children, old men . . . everybody would be fair game.

Invariably the response would be the same. I'd get cries of "Oh, no, Judge! Me and my boys, there'd be hell to pay." Or "Nobody comes into my 'hood and does that!" They'd be all up in arms about it, and up in my face for making them even consider such a scene, but when they'd finally calm down I'd tell them that what they were doing in choosing to sell drugs was no different. They were killing their neighbors the same as these phantom men on my imagined flatbed. I'd tell them how they were putting poison into their communities every day and how that poison didn't make any distinctions. For a time, I used to send kids up to the neonatal unit at one of our area hospitals and force them to look at these tiny little babies, not much bigger than an adult hand, struggling for life because their mothers took that poison while they were pregnant. (I had to give up on this tack after hospital administrators determined that their patients should not be used for such "scared-straight" visits, but for a while it was a very effective tool.)

Once in a while, a wiseass would argue the point and

tell me how the little old lady down the street didn't have the first thing to worry about in terms of drugs, but of course he'd be dead wrong. No, these senior citizens weren't doing drugs, but their neighborhoods were no longer safe on the back of all this drug activity. Addicts would break into their homes, or roll them on the street, just to get money to buy drugs, or there'd be a drive-by shooting over a deal gone wrong. The older folk were virtually prisoners in their own homes—and even their own homes weren't anything like the safe havens they deserved.

A few of these kids heard what I had to say, but too many of them didn't. Too many of them spouted off their too easy answers: *Judge, if I don't do it, somebody else is gonna do it.* Or *I never meant to hurt nobody's baby, but if the mother is pregnant she shouldn't have taken those drugs; it's not on me, it's on her.* I could argue the point all day long, but there was always another case waiting to be heard, and there's only so much preaching a kid's willing to hear from someone in authority—even if that someone is a pretty cool judge with no agenda beyond setting him right.

Take the easy-money foolishness off the streets and place it into our more affluent homes and you've still got the same mess. I heard a case in my television courtroom that set my antennae on full alert. A young boy from a well-to-do family came in on charges of stealing from his own father. The kid was barely a teenager, and he'd run a clever credit card scam in an effort to impress his friends and win favor. He was going over the household mail each afternoon before his father came home from work, and he kept seeing all these credit card solicitations from various banks, offering low rates and whatnot. We've all been on the receiving end of these letters: *You've been preapproved!* Well, this kid got it in his head to respond in his dad's name, and within a couple of weeks he had himself a valid credit card—and his father didn't know the first thing about it. Soon enough, the bills started to come in, and the kid had to figure a way to keep his scheme afloat, so he applied for another card to pay off the first one, and in this way he dug himself deeper and deeper into trouble.

Now it wasn't like this kid was running out and buying major electronic equipment or fencing high-end items on the street, which didn't excuse his activity so much as it helped me to understand it. Most of the charges were to local movie theater complexes, for four or five tickets at a time, and movie theater concessions, and occasionally there'd be a video game or two on the bill. He was using the money to buy himself friends, as a way to get himself with the in crowd, and his "friends" were only too happy to play along. The back-story to all this was that the kid had a serious weight problem and had long been ostracized by his peers, and he'd apparently seized on this false generosity as a way out of his situation. He wasn't buying stuff for himself but for these other kids. And he pulled it off, for a time. How he managed to go up to a ticket window time and time again and produce a credit card without being questioned by a manager is beyond me, but I suppose that's for a separate discussion.

Eventually, though, the kid's good fortune ran out. He

couldn't generate these new credit cards fast enough and in desperation figured a way to access his father's savings account to pay off his charges, and this was how his father finally figured out something was going on. The numbers on the man's monthly statements didn't quite add up, and the disparities pointed directly to his kid. In all, he'd rung up about eight hundred dollars in charges before his father put an end to it.

The father sought help from our show and hoped to resolve the matter on the air, which made for some compelling television and gave me the chance to extend a hand to a kid in trouble. I chose to connect the boy's actions to the negative self-esteem issues flowing from his obesity, and in my disposition ordered that he attend a camp for overweight children to help him get a handle on his eating and his inactivity. It wasn't exactly a case of the punishment fitting the crime, but it made sense to get to the heart of the matter, especially since this wasn't a criminal case in any traditional sense. Call it fraud, call it embezzlement, call it plain old stealing . . . a prosecutor could have made a case on all counts, but in my role as a television judge it fell to me to seek a hopeful resolution alongside justice. And here I felt I had to find a way for this father and son to patch over this one big issue and to make the kind of repairs they'd need to keep something like this from ever happening again. I also ordered the boy to reimburse his father for every last dime he spent on the scam, to take on odd jobs and forfeit all birthday and Christmas gifts until his debt was cleared. What he did was wrong, and he needed to get that message loud and clear, but he also needed to hear that his father loved him and supported him and was willing to work with him to ease some of the conflict that had pushed him to this place.

I was lucky here in that an easy solution presented itself, but I know it doesn't always shake out in this way. In this case, there was a transparent cry for help. In this case, no one was hurt. In this case, the transgression was a hip-pocket crime committed against a victim who didn't want to see the defendant pay too dearly. In other cases, though, the scenario plays out differently. I remember one case that fairly shocked the community when it made the papers. A young student had been attending a prestigious university. His parents assumed he was living in one of the dorms; they'd visited him there on many occasions and called him at his number there several times each week. But their son was living a dangerous double life, selling marijuana and cocaine to his fellow students; apparently he was making so much money dealing drugs that he'd moved into an exclusive condominium in a nearby gated community, without ever telling his parents or university officials he had moved. He had his calls forwarded to his lavish new pad and figured that no one in any position of authority would ever have to know. Now this troubles me, as I hope it does you. This young man had enjoyed a great background. His parents were pillars of their community. He'd never wanted for a single thing, and here he was, dealing drugs and dodging trouble so that he could earn enough cash to keep living large, and driving a nice car, and so on. When the case broke, it turned out he was hardly even attending his classes, because the classes had become secondary. School was out, and the easy money was in. The case never fell to me, because the young man was tried as an adult and put away for a good long while, but I set it out here as a caution to loving, committed parents who've managed to provide everything for their kids: You never know what your child will do for a buck, especially if that money has no real meaning attached to it.

Sometimes parents don't even know what's going on under their own roof. I struggled with one particular case involving a teenage boy stealing money from his grandmother. The boy's parents were no good; they were each serving time on separate drug dealing charges, so right out of the gate the kid didn't exactly have the best role models on which to pattern his behavior. He might have, though, if he'd looked to his grandmother. This good woman had stepped in to keep the child at home, and in the beginning she overcompensated, spoiling him with everything he wanted, handing over the keys to her car whenever he asked, whether or not she needed it to get to work. Over time she realized she couldn't sustain that kind of freewheeling attitude; the economics just didn't work, and the logistics weren't great either, so she started to pull back. She started to say no. And the boy couldn't hear it. So he started stealing from her, brazenly fisting the money right out of her purse. But what brought him into court was that he started beating on her as well. He was angry and felt entitled, and his grandmother stood in his path, so he pushed her aside in what ways he could.

"Yeah, I beat her," he admitted in my courtroom. "She started fussing at me and I just snapped."

I guess he meant this as some sort of defense, but what the hell kind of defense is that? "You want to see me go ballistic in here," I railed, "then you come in and tell me you've been beating up on your grandmother. How dare you? But for her, you'd be in a shelter somewhere, and you're telling me you just snapped?"

The boy had the nastiest, surliest attitude, and as I worked to understand his good, noble grandmother, shouldering her grandson's blows for standing up to him, I kept thinking, This is how it sometimes goes. This is what happens when our kids are out of control, when they think they're entitled to everything, that the world owes them a new gold chain. They want what they want when they want it, and pity the poor soul who stands in their way—even if it happens to be their own grandmother.

Let me return to that sixteen-year-old drug dealer, the girl who raked in as much as one thousand dollars a day before she was scared straight by her boyfriend's execution, because there are lessons to be learned from her extreme example. This too is how it sometimes goes—that we can take the terrible mistakes of others and learn from them. Chief among these lessons in this one case is that the easy money mentality that permeates our gangs and our school yards is seriously flawed—and fraught with danger. Kids see this after a time, but the trick comes in getting that message across before it's too late. They need to check it out for themselves, take it in on their own, watch a friend take a fall from their own front row seat. Parents and teachers and guidance counselors (and judges!) can sell it every which way, but a lot of these kids just aren't buying, not until they know it firsthand. Be assured: Eventually most kids get it. Eventually most kids tire of looking over their shoulder all the time. Eventually their luck runs out and they're left to face a mountain of trouble. Eventually something happens to shake them from their complacency.

These kids put themselves at tremendous, constant risk, and the hard truth is that most of them only consider the legal risks before throwing in on this front. Clearly, the bigger risks come on the street—or at least the more *likely* risks. Law enforcement officials do what they can to police the problem, but there's also a street-level form of justice to consider. I've seen kids killed in drug deals gone bad. I've seen kids jumped for their drugs or their drug money. I've seen kids fire their own weapons for no good reason but to protect their investment. After all, if you're dealing drugs, it follows that you're walking around with wads of cash, and you've probably got a weapon to protect yourself. You're a ridiculously easy target. Think about it. What are you going to do, call the cops? Plus, nobody just takes your money and runs anymore; kids have access to all kinds of weapons and think nothing of using them; they'll shoot you as soon as look at you, and money is at the bottom of most of this mess. There might be a host of social ills piled up on top, but the easy money mentality is somewhere deep in that pile. Tell me it doesn't follow that the number of violent cases involving our youth and handguns directly correlates to the upsurge in the use of crack cocaine. Fists used to rule the streets, and a broken nose or a cracked rib were about the worst you could expect from a beating, but those days are long gone; nowadays kids pull out their weapons and fight to the death, and the deaths hardly register, but the message to our kids has got to be that sooner or later a gun will be pointed at them. Oh, yes, it will. Go far enough down that wrong road and the weapons will find you.

The interesting piece to this drugs-and-money dilemma is that most of the dealers I've encountered won't touch the stuff themselves. They know what time it is. They'll say, "Aw, no, Judge. I don't do drugs. Are you kidding?" They've seen their older brothers or their uncles or their cousins caught in the grip of crack or heroin, and they're smart enough to turn away. They're not smart enough to keep from dealing, but I suppose it says something that they don't use the stuff too. Rarely, I'd come across an addict who was hustling to keep ahead of his habit, to pay for inventory he should have sold instead of snorted, but for the most part it's these easy money schemers looking to make a buck off someone else's addiction. Of course, almost every one of these kids smokes marijuana, but kids today don't consider marijuana a drug. They just don't. They'll be taken into custody, and marijuana will turn up in their systems, and they'll swear up and down that this doesn't count.

"I'll smoke marijuana," they'll say, "but I don't do drugs." And in their minds they don't.

"That's just weed," they'll say, "that's not drugs."

Drugs and money, they go hand in hand. They just do. The whole culture is built on the backs of one kid trying to take advantage of another and on neither one of them having any real sense of the value of a dollar or the work ethic that should be behind it.

One thousand dollars a day. Tax free. How do you tell a sixteen-year-old to take a job folding sweaters at the Gap when she's looking at that kind of money? How can parents hold fast against the instant-gratification, damn-the-consequences impulses of our children? The best course—indeed, the *only* course—is to turn it back on the child. Ask them what they want out of life. Ask them to tell you their hopes and dreams. Ask them to walk you through their next ten years. You'll hear they want to be a rapper, or a basketball player, or a lawyer, and you'll counter with the message that the money they're looking at making now, as much as it is, is just a little bit compared to what they can earn in success if they stick to the proper path. "None of this is going to happen if you're in jail," I tell the kids in my court.

I've struggled with this issue for years, and what I've come to realize is that none of these kids think they're going to live past twenty-five anyhow. There's a live-fast, get-it-while-you-can mentality out there, and it's pretty much impossible to get kids to think about their long-term future when so much of their world is so immediate. This is true in our worst neighborhoods and in our best neighborhoods—in our broken homes and in our tight-knit families. Every day in my courtroom there were good kids, with involved, caring parents, hauled in on this or that drug charge. Student leaders. Student athletes. ROTC candidates. Each one peddling drugs for the quick cash they might have earned washing cars or scooping ice cream. And what will turn these kids around, in almost every case, is an in-their-face message that there's more downside than upside to whatever it is they're doing. They need a dashed dream, *somebody else's* dashed dream, in order to salvage their own.

A lot of times I'll drive by a high school and see a car smashed to pieces and hauled out to a prominent spot on campus, where it stands as a vividly potent reminder of what can happen when kids drink and drive. Usually the car has been towed from the scene of a fatal crash involving students at that particular school, and it's proven to be an effective way to cut down on that kind of activity. Hell, if I were a teenager and I had to look at that smashed-up car on the way in to school each day, I'd think twice about doing ninety miles an hour on my way home from some party.

Find that dashed dream and make it resonate for your child. Find the kid in your community who had everything to lose, who lost in all with a single bad decision. The heavily recruited varsity basketball player who bet his future on some quick cash in a dope deal, who ended up losing his chance to play in college just to make an extra couple hundred bucks. That political science whiz who dreamed of going off to law school to become the next Johnnie Cochrane? The easy money he could make in a small drug deal looks mighty insignificant next to all those million-dollar book advances or those six-figure legal fees, doesn't it?

Sadly, we don't have to look too far for these *hey, this could be you* stories. They're all around. Remember Len Bias, the All-American basketball star from the University of Maryland, drafted second overall in the 1986 NBA draft by the Boston Celtics, only to collapse in his dormitory room just a couple of days later after overdosing on cocaine? It was a huge story at the time, on and off the sports pages. Man, that message hit home for a whole lot of kids—a different message from the one I'm looking to put out here but close enough so you get the idea. Indeed, Len Bias's mother recognized her son's tragic death for what it was, and for a time she took up a wonderfully courageous crusade, traveling the country urging kids to consider the devastating perils of drugs. Good for her. By all accounts, Len Bias had never used drugs until the night they killed him, and I suppose he fell in with some bad folk who coaxed him into a situation he didn't want and yet felt he couldn't avoid, and the powerful message here is that if such as this could happen to such as him, then we all better keep on our toes. Here was this kid who had everything to live for, who was living a lot of kids' dreams, who was on the very cusp of making it big, and it was all gone in a sad, foolish instant.

And while I'm on the subject of money, let me slip in a word of caution: Don't fall into the trap of substituting your money for your time. Don't slip your kid an extra twenty every time you feel guilty for missing one of her school plays. Don't let an undeserved present buy you an extra round of golf after you'd promised your kid you'd help her with her homework. And—this most of all!—don't become one of those parents who drop their kids off at the mall with a credit card and no restrictions. These malls are little petri dishes for cultivating disaster. You've got kids hooking up with kids they don't really know, getting into all kinds of things they didn't plan on getting into. Have I ever dropped my kids off at the mall? Yes, I have. Many times. But it was never the default activity for their weekend afternoons. From time to time, if they had something specific they wanted to shop for, some group of friends they wanted to hang with in an unsupervised setting, I'd let them go, but we never made a habit of it, and they certainly never went with my credit card. If you want to spend your money on your kids, that's fine, but spend it yourself, with them. If you let your credit card go someplace without you, there's a good chance you're doing so for the wrong reasons—and an even better chance that you'll regret it.

One of the most outrageous abuses of this kind I ever heard was a deprivation case involving a globe-trotting set of parents and a couple of left-behind credit cards. A high school junior from Atlanta's affluent North Side was left at home to fend for himself while his parents went off to Europe for a few months. Apparently he was the youngest child, and his parents had reached the age when they wanted to travel and the stage in their life when they didn't want to be tied down to their routines, and they no longer saw the merits of staying home and actually *raising* the children they had brought into this world. They stocked the fridge, left behind sufficient funds and credit cards, and gassed up the family car so Junior could drive himself to school and run errands. They had it all covered, except they hadn't counted on their kid running into trouble with the law. Big trouble. There were parties at the house, and complaining neighbors, and property damage of various kinds. There was also a drug possession charge that came as a result of a neighbor's complaint.

The police brought the kid in to be processed and the parents were nowhere to be found.

"Judge," the caseworker reported, "the boy says his parents are in Europe."

"How long have they been in Europe?" I wondered.

"Well, a couple of months," the caseworker responded, "from what I hear."

I thought, A couple of months? What kind of parents leave their minor child home alone for a couple of months? I don't care how responsible this young man may or may not have been (and by his actions, he had clearly tilted the scales the wrong way), you can't up and leave a child for that stretch of time. You just can't. Can you leave a teenage child for an afternoon or an evening? Sure. Can you leave him overnight? Sure. In fact, it's probably an important show of trust, to leave a responsible child in charge every once in a while, in age-appropriate ways, for age-appropriate periods. But to disappear for a couple of months? To leave a high school kid on his own in a big, sprawling house with unlimited funds and no checks and balances? It was unconscionable.

As the case played out in court, I became less and less concerned with the drug and nuisance charges against this kid and more and more concerned with the deprivation issue created by the parents. Yes, there are a lot of ways you can deprive a child, and this child of affluence, left to fend for himself with his parents' money and car and credit cards, was very clearly deprived. Abandoned and deprived. He wasn't hungry, and he didn't want for any material things, but he didn't have any kind of parental supervision, and as we assessed the history it became clear that he hadn't had much in the way of parental supervision for the longest time. His parents were guilty in an extreme way of trading their money for their time, so I sentenced the whole lot of them to family counseling so they could get their shared act together, and reported the parents to authorities on charges of child neglect. Oh, there was some community service aspect to my sentence for the child, just to keep him honest, but I felt strongly that the family counseling was the key to turning the situation around.

In closing, I'd like to bring the conversation back under my own roof—because honestly, it's too frustrating shining such an intense light on other parents' missteps. Better to look at the positives where you can find them, don't you think? Happily, in our house, I've never wanted for positives. My kids are turning out okay, and I must credit each of them for that, and in this one regard I also credit our shared interest in understanding what money means, what money buys, and why money matters. It's been key.

Earning money and learning to manage it responsibly have everything to do with setting realistic goals and understanding pragmatic limitations, and Chris and Charles have each made strides in this area. Charles was always good about taking on part-time jobs, but Chris has had to scramble because of his athletic commitments. (He participates competitively in four different sports, including a top-flight weight-lifting program that eats up much of his summer.) My thinking on this is that it's okay for kids to put off finding a job if they have pledged their after-school time to a worthwhile pursuit. Team sports, individual sports, enrichment programs of various kinds . . . it's all good, and it's all important. I've known parents who've compensated their dedicated children who can't find the time to hold down a job because of all of their extracurricular activities—and I'm all for it, if that's what it takes to make the situation work in your household. I can certainly see putting a kid on an allowance-salary for all the time put in after school working for the school paper, singing and dancing in the school play, or tossing a football for the varsity squad, especially if the child needs to earn extra money and can't find the time to keep a job with all of these other commitments. There is a kind of work ethic in focusing time and attention on a team practice schedule, the same as there is in mopping the floor down at the local factory, and if a parent has the ability to support that ethic in kind, then I say go for it.

Even volunteer initiatives should be championed over menial-labor type jobs. There's something to celebrate in the child who gives freely of his or her own time, wouldn't you agree? The spirit of volunteerism is essential to our shared well-being, and though on a personal level I didn't underwrite my kids' social action efforts, I won't second-guess parents on this score. Heck, if that's what it takes to get your kid to volunteer at the local soup kitchen, then that's what it takes. With my kids, they understood early on that from those to whom much has been given, much is expected, and when they were younger we used to get our heads together to come up with community-based projects where we might all lend a hand.

Teaching your children to be responsible about money doesn't end with earning it or managing it—they've got to understand how to invest it, as well. Every month, as part of my children's allowance, I'd put fifty dollars into a mutual fund account for each of them, hoping to instill the message that part of what they earned they needed to save, and as the funds grew their interest in investing kept pace. Chris in particular became fascinated with the stock market, to the point where he even opened his own trading account with one of those low-commission on-line brokers. At eleven, Chris was researching companies on the Internet and seeking out opportunities on his own. If he found a company he liked, he might buy a couple of hundred dollars worth of stock with the money he'd saved from birthday and Christmas presents and such, and after a while his portfolio really took off. Looking back, I suppose anything he touched might have turned to gold during those heady times in the market, but back then we all thought he really had a knack. And maybe he did. He found one software company that was developing a method to translate voice commands into a printed manuscript, and he became real excited about it; happily, other investors became excited about it too, and the stock took off.

At some point Charles took notice of his little brother's success in this area and asked for his help. He set up his own on-line account and set aside his own savings for this purpose. "Whatever you buy for yourself, I want you to buy for me," he told Chris, and Chris was happy to oblige. Well, within a short time Chris had nearly doubled his brother's investment, and I turned to Chris one afternoon and inquired about his arrangement with his brother. I didn't want to stir up trouble, but I thought it might be a good opportunity to get Chris to realize that he was providing a valuable service that should not go unnoticed. "Your brother ought to be paying you to manage his portfolio," I suggested to Chris at one point in our conversation. "Really, you're doing all the work. You're doing all this great research. He's not doing a thing."

At this, Chris turned to me and said, quite reasonably, "Mom, you don't get how this works. If I make money for

Charles, that's cool. If I lose his money, he's gonna kick my ass."

When I stopped laughing, I realized that in this exchange was the application of many of the lessons about money I'd hoped to pass on to my children—and a few lessons I'd never even considered.

## *Have a Little Faith*

I've spent a lifetime taking things on faith, even faith itself, but at the same time I've had my share of deep, troubling uncertainty—never more so than when I lost my dad. My father was a wonderful, wonderful man, and he and I enjoyed a wonderful, wonderful relationship. I don't mean to sell short my relationship with my mother, but even she will tell you there was something special going on between me and my father. I'm very close with my mother, don't misunderstand, and she's been a fiercely proud inspiration who has loved me and nurtured me through all the difficult and joyous times in my life. But I was always closest with my dad. He was a great parent, a great friend, a great counselor. From time to time I used to sneak away from the courthouse to join him for lunch back home in the house I grew up in, and in those few rushed moments over our kitchen table, between bites of this or that thrown-together sandwich, he'd set me right and clear my head and send me back for my afternoon calendar thinking I could meet any challenge the system threw at me. If I was weighing a particularly difficult disposition, he'd encourage me to pray on it. And I would. And in the praying, I'd see my way to a resolution, and but for the praying I might not have discovered the same course.

The day my father died is stitched painfully to my soul. I'd been out at an early dental appointment and my secretary took the call. The moment I drove up, I saw my secretary waiting for me outside the courthouse and I knew something was wrong. Most working mothers would process a scene like that and think maybe there was a problem with one of their children, but I knew instinctively it was my father. I just knew. He'd been in great health—he'd played in a charity golf tournament just the week before—but my gut told me something had happened to him.

My secretary waved me down, and I pulled up to the front entrance to see what the matter was. She jumped into the front passenger seat. "Your dad's been rushed to the hospital," she said.

She told me what hospital he'd been taken to. If I'd been thinking clearly, I'd have flagged down the sheriff and had him take me in a patrol car, with the lights and the wailing, but I wasn't thinking clearly. I wasn't thinking at all. I pulled a U-turn in the parking lot and sped off, and by the time I reached the emergency room my mother was sitting in the waiting area, her chair backed against the wall, a man I didn't recognize sitting next to her offering what comfort he could. My mother is a stoic woman, strong as a rock in every situation I'd ever seen her in, and here she was, crying. I thought, What in the world is going on? My brother Paul was there too, but I didn't stop to get their report. I made straight for the main nurse's station, corralled the physician in charge, and pushed my way through to the room where they were working on my father. I wasn't supposed to be there, but this was my dad, and I wasn't about to sit quietly out in that waiting room while he lay suffering across the hall.

There was an oxygen mask on his face. He couldn't talk. I reached for his hand, and he reached back. Anyway, he returned my grip, with strength. "Dad," I said, "it's gonna be all right."

I believed this with all my heart, so there was nothing else to say, and I stood there for a few beats longer, holding his hand. It was just a moment, but in memory it lingers as the longest while, until finally the nurses pulled me from his side and directed me back to the waiting room.

I tried to sit with my mother and brother, but I couldn't sit still. I heard one of the nurses call out a Code Blue, but I didn't know it was for my dad. They'd pulled the curtains, and I could no longer see him from my spot in the waiting room, and I guess I tuned out the noise of urgent business all around. My mother filled me in, and she kept saying she didn't think Dad was going to make it. She told me what the doctors had told her, but I couldn't hear it. I couldn't accept that he was dying. I said, "What are you talking about? Of course he's gonna make it." But of course, he would not.

I went headlong into a take-charge mode. It must have been some sort of coping mechanism that left me thinking I could fix the situation with a couple of phone calls, and there I was, working my cell phone, trying to line up the best doctors in the city, trying to bring in enough second opinions so that one might go a better way. Insurance? That was not one of my worries, at just that moment. Dad was covered, I knew, by every cross and shield you could imagine—in both the health care and the spiritual realm. I tried to track down a dear lifelong friend on the medical faculty at Emory, Dr. Roderick Pettigrew, thinking I could urge him to put a team together to take over my dad's care, and all the while my mother and brother were consoling each other at the other end of the room. We were just ten or twenty feet apart, but we were in a completely different place.

After a while, one of the doctors came out to talk to us. By this time we had been joined by our pastor. My younger brother, Kolen, and his wife, Jeannie, were on their way. The doctor told my mother the situation. He told her my father had been on a respirator for more than thirty minutes and that there was no longer any hope of saving him. About the best we could hope for, he said, was for the respirator to keep him alive. He put it to us plain: "At this point, we are standing in the way of God taking him home."

My mother thought about it for a heartbeat, and only for a heartbeat, while I think about it still. "Then let God take him," she said, and the words spilled from her lips with certainty, purpose, dignity, and a deep, abiding faith.

I was in shock. My mother and brother had had those long moments together in the waiting room, and they were steeled for this turn, but I wasn't prepared. Not at all. They were in a place of acceptance, and I was toiling in denial, still firm in my belief that there was something I could do, someone I could call to set things right. I thought, This isn't happening. This can't be happening. My father was seventy-two years old, but he could have passed for sixty. He didn't smoke. He was in tremendous shape. He played golf, horsed around with my kids, never sat still. It was a Tuesday, and he'd just taken my kids to a football game that weekend in the Georgia Dome, which was still being talked about that morning at breakfast as a great big deal.

Let me tell you, this was a real test of my faith. Ask me any day leading up to that one whether or not I believed in God and I would have answered a resounding "Yes!" Ask me any day since my father's funeral, and I'd once again answer in the affirmative, without hesitation. But on that day, at that moment, I didn't know what to believe. I was angry. I was confused. And I was in pain. For some reason, I thought in terms of what I did for a living, sitting in court all day long, listening to all those people who abused their children, who neglected their children, who abandoned their children, and I railed that those folk were still here and my dad was gone. It all seemed so terribly unfair, so arbitrary. I thought about all those absent fathers, and I thought about my own father, who had been very much present in the lives of each of his children and his grandchildren, and there seemed to be no justice in it. None.

It was a real turning point for me, and what turned me around, back to my place of faith, was the funeral. Actually, it was the funeral, where I'd given myself a job to do for my dad, mixed together with my uncertainty. I was at a crossroads and I put it to myself plain: either I believed, or I didn't. My dad had always talked about standing at the corner of a place he'd called New Hope Road, about moving ahead without looking back, until you were far enough along on your more certain path that the road you'd left behind was no longer visible. Either I accepted my father's fate, or I didn't. Either I had faith in a divine order and a higher being, or I didn't. Either I could move on to the next phase in my life and the life of my family, or I couldn't. I stripped it down to the bare bones and what came back to me—slowly, eventually—was a reaffirmation of that faith. It didn't happen straightaway but through an awful lot of prayer and soul-searching and agonizing reappraisal of the values I'd always held dear. At the time of the funeral, though, I was still so deep in doubt that I didn't think I'd ever claw my way out of it. I was planning to speak, because I felt no one else was in a position to say the things I thought needed saying, and yet if I couldn't summon the strength to accept the condolences of family and friends on a one-to-one basis, how did I expect to speak my heart to hundreds of people at the funeral? How could I do my father the honor he so richly deserved? In many ways I was in no shape to speak, but there I was planning to offer a eulogy. My mother and brothers wondered if I could get through it, and of course I was wondering the same.

Finally, to help me through, my mother offered her favorite words of encouragement and an affirmation of her faith—both in me and in the Lord above. "God is sufficient," she whispered.

Just then, in my darkest, most despairing moment, just before the service was due to begin, I was overcome. Reinforced. Swept away by a cool breeze that left me grounded and centered in ways I hadn't been before. It's impossible to describe, really, but there it was, and there I was in the face of it, fortified with a strength that was not my own. And with that strength came the reaffirmation I'd been seeking. I believed once more. I was angry with God for taking my father so soon, so swiftly, and there are moments when I am angry still, but I no longer question it. I have been returned to this place of faith and acceptance, and I plan on staying here, because really, I don't know how you heal from the pain of a loss such as this without that extra force outside yourself, without God's amazing grace.

And so I stood and said my few words and in so doing turned that corner my father had always talked about, down the path of New Hope Road. Toward hope. Toward acceptance. Toward faith.

## Simple Strategy #7

# Reach, Teach, and Preach

I come from a long line of churchgoing folk. It's a part of who I've always been and a part of who my children will become, and in fundamental ways it's been a part of every decision I've ever made from the bench. And yet I'm here to tell you that even from this faith-based perspective there's been room for doubt, while beneath that room for doubt there have been opportunities for growth and insight. A little bit of doubt can be a good thing, wouldn't you agree? Especially when it gives you the chance to try on a new worldview.

Now I know there are those in this world for whom faith is irrelevant or beside the point. They go through the motions of believing, or they don't bother, or it's not even a part of their frame of reference. Or they consign faith to foolishness and prefer to take their chances on their own. Understand, when I talk about faith in this way, I don't mean to limit the discussion to the God-fearing, God-praising folk in our churches and mosques and synagogues; I throw into this mix everyone who has a belief in something bigger than herself, everyone who recognizes some kind of order in the world, some grand system of checks and balances. That belief can come in the context of relationships, of family, of community, of humanity, but it must come. It must. In one way or another we must be accountable, and we must make our children accountable, and it's a matter of personal choice whether that accountability stretches to some higher being or to some vaguely delineated sense of civic-mindedness or to simply doing the right thing by the right people. But make no mistake—it must come.

For me, in my family, that accountability has been to God. I'm so clear in it now, it is the anchor that sustains me. I could not have stayed on the bench all those years without the strength I took from this firm belief, the strength to move from grieving about all those lost, untethered children to helping them chart a path for the rest of their lives. I could not have gotten past my father's death. I recognize that there are people who are not anchored in quite this way, but I don't know how they do it. And I don't know how they can ask their children to do the same.

I start every day in prayer—for my children, for my family and friends, for my community, for the world around. Indeed, not a day goes by that I don't pray for my children, and it's always a dual offering. It's a prayer of thanksgiving, because I'm truly grateful for the wonderful blessings they've become, but it's also a prayer of safekeeping, so that God will keep his hands on them and keep them in his care. Lord knows we parents need all the help we can get.

These days I'm back attending the church I grew up in—the Providence Missionary Baptist Church in Atlanta. It could be a Methodist church, for all I care; the point is that it's a part of me, and my family is a part of it; it's where we belong. My parents joined the congregation after they were married in 1949. I was christened in that church, and baptized in that church, and a lot of the old folk are still around. And that's the best part of the whole deal—the fellowship that comes from belonging, from throwing in with a group of like-minded folk, from the shared history and perspective we've managed to develop over the years. I don't feel as if I go to church to meet God, because I believe God is always with me, but I go to check in with the rest of my community, to keep myself grounded and whole. These are the people I grew up with. My kids always tease me about it. They tell me that when we go around to do the offering I always break the line when I go to meet and greet the other congregants, and I'm always the last person to sit back down, and it's true. I am. I'm busy hugging Mama Hannah, a graceful and joyful soul who at ninety-three has become a surrogate grandmother to me and great-grandmother to my children; or there's Deacon Moore, a sweet old man who at ninety-two is more beacon than deacon. I'm surrounded by a congregation of people who have known me all my life, and I'm not about to sit back down until I get all my hugs in.

And happily, it hasn't been like pulling teeth to get my kids to go to church. They've made real connections of their own—most especially with our pastor, my dear friend Gerald Durley. But sometimes, in some families, it *is* like pulling teeth, and it falls to us parents to offer our children the guidance they'll need to steer them down their proper paths. If it's not church—whatever your religion—then it needs to be something else, some other forum in which we can foster responsibility and respect. I've seen too many kids in my courtroom who don't have the first idea that they're part of something bigger than just themselves. Absolutely, kids need to understand that there is a force greater than they are. Whatever you choose to call that force—Jehovah, God, community, the greater good—it needs to be acknowledged and encouraged and held out in such a way that our children can depend on it or look to it for inspiration or comfort or guidance.

At bottom, it's the faith we have in ourselves that propels us forward, and it's our willingness to learn from our missteps that allows us to reach beyond what we ever thought possible. Sadly, the best way to illustrate this basic truth is in the negative and sadder still is the fact that in my courtroom there were enough illustrations to test anybody's faith. A few of these stood out, like flash points, and in this way stood for every textbook bad decision that seemed rooted in faithlessness and thoughtlessness and heartlessness. Consider: There was the blank, remorseless face of a teenage assassin hired by a drug dealer to bump off another drug dealer for a thousand-dollar payoff. This one has stayed with me, I think, for the complete immorality of the act and the brazen disregard for society it represented. The defendant was sixteen years old, and he didn't seem to have a clue. The only values that seemed to mean a thing to this kid were the interconnected needs to make a buck and cover his own butt. The guy he was hired to shoot was in the hospital at the time the hit was taken out, so this kid marched into his target's hospital room, pulled out a small-caliber handgun, held it to the drug dealer's head, and pulled the trigger. Just like that. No thought for the life he meant to take. No thought for the mess he was making for himself. No thought at all. Trouble was, the kid didn't know enough about human anatomy to place the gun at his target's temple, so he shot him through the cheek instead. The victim was right there in the hospital, and the staff were able to tend to him right away and apprehend the kid who shot him, and the hit went wrong in so many different ways it's a wonder it nearly came off as planned. The assailant was charged with attempted murder, and he came before me in court with absolutely no regard for what he had done—or for what he had tried to do. The decision to go out and kill somebody held about as much weight with this kid as the decision to change shirts. I couldn't understand it. There was no regard for human life, no remorse, no feeling whatsoever, and I remember thinking that the true evil here lay in the way this young man had been raised.

Even more horrifying, if you can even quantify such things, was the case of a young girl repeatedly raped and eventually impregnated by her own father, who for some inexplicable reason refused to testify against him and even refuted the charges against him. What happened was that the girl's mother had died. At fifteen this young girl was the oldest of several children, and the father began having sexual relations with her. It was never clear if he had been abusing her while the mother was still alive, but there was no denying the abuse from this point forward. One of the younger siblings eventually confided to her teacher what was going on at home, and the teacher reported the situation to the Department of Family and Children's Services, and it fell to me to sanction the removal of the children from the father's home.

The oldest daughter, whose name was Tanisha, took the position that she had become pregnant when she was raped in the front seat of a car, but she kept doubling back on her own story. First it was a two-seater sports car and then it was a sedan with a front-seat bench. First it was an acquaintance who had raped her and then it was a complete stranger. First she claimed she could identify her assailant in a lineup and then she maintained that it had been too dark to see clearly. The DNA evidence later confirmed that her father had in fact impregnated her, and there were damaging accounts from some of Tanisha's siblings, but Tanisha wasn't changing her story, and neither was the father. As I listened to the whole sorry, sordid mess I couldn't help but wonder where faith had been in that household, or how it was that a father could do such a thing and still not acknowledge that he had done something wrong, even in the face of such overwhelming evidence. It confirmed for me that where there is no room for faith there is no hope for healing, and that when there is no remorse there is no chance for rehabilitation, and I moved to do something about it.

Understand, the matter before me was confined to the department's request for custody of these children, and this piece was clear. Absolutely, the kids needed to be out of that house, immediately. But there were so many other pieces flowing from this one motion and so many opportunities for this empty, hollow man to admit to what he'd done and begin to make emotional repairs that I couldn't look away. Not only did he impregnate his own daughter, but he refused to get her any prenatal care and delivered the child himself at home! It wasn't just one unconscionable act but a series of unconscionable acts strung together into an atrocity, and it was this thorough and persistent lack of conscience and accountability that set me reeling. I ended up trying that case until about one o'clock in the morning, because I was afraid that if I adjourned I'd never see the man in my court again. He was there to fight the Department of Family and Children's Services motion for a transfer of custody, and there were no criminal charges against him, but here I'd been presented with overwhelming evidence to support sexual abuse and child endangerment charges—enough, certainly, to bind this despicable man over to the grand jury. So that's what I did.

It took until the wee hours to get all the paperwork in order, and we had to call in extra security to keep the man in check until we got through the process, but we got it done, and the message here is that in the absence of any kind of moral compass it falls to the court to set things right, no matter what.

You might wonder what these cases have to do with my opening remarks on God and faith and purpose, and there was a time when I might have wondered the same, but I now maintain that they are all connected. Life happens. Mistakes happen. Faith happens. Redemption, too, happens at the heels of all three. Good kids can turn bad, and bad kids can turn worse, and it falls to those of us charged with keeping the balance to weigh the transgression against the consequence. Clearly, if there'd been a stronger moral compass at home, these people might not have wound up in my courtroom, but what's also clear is that there was no changing the outcome of either of these two cases. I couldn't undo the not-so-sharp shooting of that sixteen-year-old assassin any more than I could the delivery of Tanisha's child by her father's hand—but I was looking for a different sort of consequence. I was looking for a measure of human feeling and compassion that seemed in some way changed by the wrongdoing itself. I was looking for a kind of remorse—or at the very least, an understanding that what went down should not have gone down in quite that way, and since it had, there were some things that needed to be done to turn the situation around.

Contrast this despicable father and this cold, callow hit

man with the unwed young mother who disposes her unwanted child in a Dumpster and hopes desperately to keep her condition and her handling of it from her parents and from the authorities. It's one of the ongoing, unyielding horror stories of our time, the way so many teenage girls go through these unwanted pregnancies without any prenatal care or—worse, perhaps—any real understanding of the changes going on in their own bodies. That their own mothers and teachers and girlfriends also look away from what has to be an obvious set of physical changes is bewildering, and troubling, but it begs a whole different set of questions. Let's focus here on the pregnant teenager's naive and frightened perspective. The scenario has become a part of our social landscape, and it knows no class or station.

Indeed, there was a famous case down in Atlanta that made national headlines and stood as a confounding emblem of this phenomenon. A young girl had her baby at her high school prom. It sounds like a bad idea for a bad television movie, but this young girl got all dressed up to go to her prom, went to the girls' bathroom to deliver her full-term baby, drowned the baby in the toilet, cleaned herself up, and went back out on the dance floor to join her date as if it were nothing at all. It wasn't my case, but it got a lot of attention, and the confused girl was charged with murder, but she responded to the charge in appropriate ways. She admitted her wrong. She'd been caught in a swirl of blind panic and denial. She wept inconsolably. She sought counseling. She asked for forgiveness. She was prepared to serve her sentence and make herself whole.

I heard dozens of these kinds of cases in my time on the bench, but the difference here is that there was almost always remorse. After the fact, after the act, there was almost always remorse. There was a prospect for growth, and learning, and there was profound sadness, and lots of tears, and I made it a point to take extra time and extra care with the kids who came before me with some evidence of remorse. Show me that you truly understand the depths of your mistakes, show me even the slightest contrition, and I'll go out of my way to put a plan in place to promote healing. I'll still come down hard, but I'll come down even harder on the kids who show no regret.

Over time, I began to look for teachable moments in my courtroom, the points of pause in the hearing of a case or in its disposition where I could offer a child a leg up or a lifeline or a second chance. These moments ran from the mundane to the compelling, but in each case they were marked by a child's willingness to come clean and make amends. Lately, these moments take dramatic shape in my television courtroom, pretty much on a weekly basis. My producers call them "interventions," and I mean to intervene in life-changing ways. After all, if you're going to shake things up and set kids straight, you might as well get results, right? But back in my courtroom in Atlanta, the results were much the same. . . .

Two teenage girls had gotten into a car with a couple of guys out for a joyride. The girls were good kids, good students, had never been in any kind of trouble, but here were these cute guys offering them a ride and a good time, and they went against their better judgment and hopped in. Turned out the car was stolen. Turned out the driver was pulled over for some moving violation, the policeman ran the tags as a matter of routine, and the whole lot of them got hauled in and charged. The charge against the girls was theft by receiving stolen goods. That's how the charge reads in a lot of states if you're a passenger in a stolen vehicle; the driver and any accomplices in the act of stealing the vehicle are charged with auto theft, but what a lot of kids don't realize is that an unwitting passenger can be charged too.

The girls were distraught. Their parents were visibly upset. They were all good people. The only thing these girls were guilty of was bad judgment. Clearly, it didn't serve any purpose to lock them up, or to put them on probation, or to blemish their record in any way over such as this, but at the same time I felt there was a life lesson to be learned.

"I never want you to take for granted what you have in your life," I directed the girls from the bench. "Your parents are all here with you today. Your mothers are in tears, your dads are fighting tears, and their hearts are broken behind this. You made a bad decision that just snowballed, and now you could be looking at a mountain of trouble."

At this point, I made them turn around to face their parents, to look them in the eyes, while I continued with my speech: "I don't know what it's going to take to get things square with your parents, but that's on you. You've got to do something to make this right, to earn back their trust. But it's on me to tell you that you're never to take for granted the blessings you have in your life. You each come from loving families, loving homes. You've got all kinds of people who care about you, who support you, and you can't ever risk those relationships on a fool move like getting into a car with two boys you don't know, two boys who ultimately won't be there for you."

I'm a big believer that the punishment should fit the crime, so here I hit on the notion of forty hours of community service at a local homeless shelter, the underlying message being that these girls should see other children who don't have what they have. They had to read stories to the kids, sweep, cook, serve, be generally available to do whatever needed doing. One of the girls had a talent for doing cornrow braids, so she ended up spending a lot of time braiding the younger girls' hair, and in this way they took in the lesson that their lives were indeed a blessing and that their blessings needed to be cherished and safeguarded.

Fast-forward to a few years later. I ran into one of these girls in church. She had a relative who was a member, and she was visiting, and she sought me out after the service. "Judge," she said, "you won't remember me, but I was in your court." She recapped her case, and soon enough I remembered, and then she allowed that her time in that shelter was the best thing I could have done for her, how it made her change her perspective, and how she was now in college and doing really well.

Two years after that she was back in my church once

more, and this time I didn't need the recap. She was headed for law school in the fall, and we were hugging and crying about it, and caught somewhere in our quiet celebration was the realization that her run-in with the law had truly turned her life around. She'd turned the corner onto New Hope Road. She wasn't a bad kid, but she'd had a moment of bad judgment, and the harsh lesson is that one bad decision can have rippling effects on a young life. If I hadn't been the judge, she might have been locked up for a couple of weeks and run in with some hard-core kids, and who knows how her life would have turned out? Instead, her service in the shelter got her to think outside herself, to count her blessings, and to think about sharing them instead of throwing them away. She got to go home each night to a warm bed in a loving home, but the kids she was dealing with had no place else to go. She needed to see other kids who didn't have what she had, to learn to think specifically about the choices she was making and the ways those choices could spin out of control.

On the show, I get a chance to offer up creative sentences that meet a decidedly more positive agenda—and the hope here is that the lessons learned resonate far beyond our studio courtroom. One of our most dramatic interventions came about when I sent a kid on a field trip to a county morgue after he had stabbed his brother with a kitchen knife. The brothers were fifteen and sixteen, and they'd been arguing over a matter neither one could later recall, and the argument turned ugly enough that the older brother reached for a knife and lunged at the younger brother. I don't believe he intended to kill his brother, but he certainly meant to hurt him, and it was clear from the file we had on him that he was a gang wannabee. He'd been in and out of juvenile court, and he wouldn't go to school, and he carried the kind of terrible attitude that begged for a wake-up call. The younger brother wasn't much better, if you want to know the truth, so we arranged to send them both down to DeKalb County, where the coroner was kind enough to let them watch an autopsy.

"You need to understand that death is final," I instructed the two boys from my television bench. Then I turned to the older brother and said, "You could have just as easily caught your brother in the neck, in the jugular, and he could have bled to death before anybody ever got there."

They were a couple of hard cases, these brothers, and they were pretty much unmoved by my speechifying, but the trip to the morgue shook them up big time. The older one came running out of the room where they did the autopsy, hugging his brother, sobbing to his mother. "I'm so sorry," he kept saying. "I'm so sorry." And he was. He was also hardheaded, and belligerent, and I didn't think there was any better way to get this difficult message home.

We took some heat for that show, for putting a couple of teenagers face first into such an agonizing situation, but sometimes you have to turn up the flame a little bit to keep kids away from the destructive fire. Sometimes it takes pushing the envelope a little bit to get the message across. We took some more heat that same television season for staging a death during an intervention at a hospital emergency room, and I could almost understand it. This was a fairly emotional piece. A teenage girl had been out drinking and partying and driving, thinking it was no big deal, and we sent her to observe a typical night at a city hospital. Our producers arranged for a bunch of hospital folk to stage a little scene for this confrontational girl. A teenage girl about the same age as the defendant, her "mother," an ambulance driver, a couple of nurses and doctors, all agreed to participate in our intervention. The girl was wheeled in, with the "mother" racing behind, blood all over the place, doctors barking out orders, commotion all around. Our drunk driver was dressed in scrubs and observing the entire scene. We had it rigged so the machines went flatline, and the girl was covered up by a sheet, and it really looked as if she had died. Our drunk driver about freaked, and as a part of her healing she even wrote a letter to the "mother" of the young girl she thought had died, telling her how sorry she was. She really was shaken up by the experience. Of course, she was pissed when she found out the whole thing had been staged, but I put it to her plain. "A lot of people went to a lot of trouble to get your attention," I said to her after her night in the emergency room. "To get you to understand that that could have been you."

It may have been a little over the top, but the message took.

One of our most powerful interventions on the show involved a young girl named Candace, who turned up in my studio courtroom after having let a man fondle her breasts in exchange for three hundred dollars. Now I don't believe in my soul that that's all she did for her three hundred dollars, and I'm not entirely sure her mother believed it either. To my ear, her story sounded as if she was flirting with prostitution. If her self-esteem was so low that she would let some lowlife paw her for money, things would only get worse from there. Her mother apparently agreed and brought her daughter on the *Judge Hatchett* show to try and set her straight. Candace had already attempted suicide on two occasions, and by all appearances, outward and otherwise, she was in a bad place.

Let me tell you, it was almost impossible to look at this stunning, innocent face on the small screen and imagine some of the things her mother was alleging she had done, but the more we talked, the more I could hear the pain in this young girl's life.

I "sentenced" Candace to two nights on the streets with two former prostitutes, in the hope that she'd pull away from that lifestyle after seeing it from the perspective of two old pros. One of the former prostitutes, Pommie, was herself a beautiful woman, although not quite so young as she had been when she started turning tricks. Pommie took Candace and our camera crew to a downtown bridge where, several years earlier, a trick had beaten her up, thrown her from the bridge, and left her for dead, and Pommie told Candace what it was like to claw her way back from a moment like that. It was almost eerie the way our cameras picked up the two of them in silhouette against the streetlights, their hair pulled back in matching ponytails; they could have been sisters but for the wrong roads that still (thankfully) separated them. Pommie talked openly to Candace about the choices she'd made in her life, about how she wished she was fifteen again, about how destroyed her world had become. And then, toward the end of a good long talk, she let loose her bombshell.

"I'm HIV-positive," Pommie told Candace. "This is what I have, and I will eventually die from it."

Candace was blown away, and so were our viewers. Pommie was a powerful messenger, and it was a powerful message, that Candace needed to want something more for her life than Pommie's reality. It's not enough for me to want these things for the kids I meet in my courtroom, and it's not enough for their parents to want them on their kids' behalf; they've got to want it for themselves. They've got to find and reach for their dreams. They've got to figure out what they want to accomplish and come up with a way to meet those goals. They've got to think there's something better to live for than to have some sleazy man put his hands all over them for three hundred bucks.

Candace was so turned around by this streetlight exchange with Pommie (Candace now mentors kids at Boys and Girls Clubs after school) that she came back the following season to participate in another intervention, this time with another young girl who claimed one hundred and five sexual partners by the time she was sixteen. The point here was that I expected more from Candace than simply getting her life back together. I expected her to lift as she climbed, to become a part of the solution now that she had moved from being part of the problem. And she did, to great effect. She told this confused sixteen-year-old that she'd been in the same boat. "I didn't want to listen to the judge," Candace confided, "but let me tell you, girl, you don't know nothing."

No, she didn't—but she would know a few things soon enough.

Ultimately, though, the most effective interventions were the ones I made in my Fulton County courtroom, because these were never planned. They came from my gut and my heart, on the fly, weighed against the presenting issues of each particular case. High on my list of favorites here is one that took me back home to how it was with my own parents, laying out a code of conduct for me to follow as a small girl. Here's that story. . . .

A teenage boy—another gang wannabee—was brought in on a purse-snatching charge, and the pieces of his story didn't quite add up. The kid copped to the incident, in a broad strokes kind of way, but his account left me thinking there was still hope for his future. After making off with the victim's purse, the kid doubled back and made to return it to her, which I thought was something. Of course, the victim must have thought he was returning to do her some more harm, but that's not where this kid was coming from. A kid who takes your purse doesn't turn around and come back, no way. If he's going to hurt you, he's going to hurt you that first time, not as any kind of afterthought. Anyway, he came running back toward his victim, and she thought he was attacking her all over again, and she started screaming and carrying on.

But that wasn't how it was. In his defense, the kid turned to me and said, "Judge, I really felt bad that I'd done this thing," and I believed him. I truly did. He had no prior record. He came from a good home, raised by an attentive, elderly grandmother. His parents had some troubles, but they'd been out of the picture a good long while, and there was nothing to explain this behavior except a momentary lapse in judgment against a tremendous amount of peer pressure.

It turned out the kid was something of a nerd, and he'd gotten caught with the wrong crowd, and he'd snatched this woman's purse on a dare before thinking better of it. He'd taken the dare to be accepted; to snatch the purse was to be allowed "in." He was out to prove himself, nothing as formal as a gang initiation, but it was clearly a test of some kind, and this young man clearly wasn't up to it. Nevertheless, like a fool, he snatched the purse, realized he shouldn't have done it, attempted to give it back, and wound up in my courtroom.

I did not want to send this kid to jail—and I thought instead to send him to my mother. She worked the Sunday morning soup kitchen at our church, and I knew she'd work this young man to the bone, so I sentenced him to four consecutive Sundays feeding the homeless. We actually had a family meeting on it, because my brothers weren't too happy with me for dispatching these kids from my court to my mother's care, but I prevailed on them to believe that this was a good kid who'd merely made a bad decision and that there was no way I'd put my mother at any kind of risk.

And my mother—bless her!—she was up for it! She said, "Bring him on!" And so we did.

Sure enough, she worked his little butt off, had him report at six o'clock on Sunday mornings, and put him to work sweeping, cleaning pots and pans, setting and bussing tables . . . the whole deal. He didn't sit still but to grab a bite to eat for himself. And there was a method to my homespun madness. It wasn't just about my mother working this child to death but about putting him alongside a bunch of folk whose lives hadn't quite turned out in the ways they'd planned. This was a teachable moment staring right back at me, and I had to think that this kid would get to talking to some of these homeless men, and he'd collect their stories and bad choices and weigh them against his own. And that's just what happened, but there was an unexpected piece that put this unconventional sentence over the top. See, my dad struck up a friendship with this young man, to the point where he was actually leaving the house at five-thirty on Sunday mornings to go and pick him up and drive him to the church in time for my mother's chores at six. And one of our deacons, Deacon Whatley, took a special interest in the kid too, so even after the kid had logged his four Sundays, he kept coming by the church to help the deacon with his Meals on Wheels program. So here we had this troubled kid with a good heart and a bad batch of peer pressure, turned around on the back of my mother's hard work and the up-close relationships he was able to forge with adult males who had come down on each side of the same bad decisions. He could see through these men what the path to righteousness might look like, as well as the path to ruin, so it turned out to be a win-win scenario, in more ways than I could have imagined.

In this case, this one act of faith in this one boy turned out to be a step in the right direction. Of course, after this the joke in our house and in my courtroom was that I was inclined toward cruel and unusual punishment, consigning these wayward children to my mother's care. She could be a tough old bird, my mother—and she had to be, as assistant principal in a high school in one of Atlanta's worst neighborhoods—but here she was, working this young man beyond weary and at the same time jacking him up and setting him straight. My brothers and I might have teased that it was cruel and unusual punishment, but it did the trick.

And so I return to my mother and father and the teachable moments of my growing up. In many ways, the firm, loving example they set established the blueprint for how I'd parent my own children and how I'd judge the children of Fulton County, Georgia. I can still recall the late afternoon when my mother called me home for dinner, and I was out riding a steep hill on my bicycle and hardheaded enough to ignore her call, and in my just-one-more-time push down the hill I hit a rock and flipped over my handlebars and tore up my knees and my arms and my face. Really, I was beat up something awful in that fall, and I walked into the house all hurt and cut and bleeding, and my mother just looked at me and said, "If you'd have come when I called you, this wouldn't have happened." There was no arguing the point. It was my mistake, and it was up to me to learn something from it.

It fell to me again, as a senior in high school, when I chose to miss curfew. God, my curfews were always so early compared with those of my friends, but my father was a stickler on this one point. And I was usually good about keeping those curfews, I really was, except this one night, when I was out in the driveway kissing some cute guy in the front seat of his Volkswagon bus, losing track of the time. I thought later I could argue my way out on a technicality. I was home, in a sense. I was on our property. I just wasn't inside the house. But my father wasn't buying it. I'd already been accepted to Mount Holyoke on a full scholarship. I was graduating with honors. I'd never given my parents a moment of real, substantive worry. I'd have thought I'd have a couple of extra legs to stand on, or at least some kind of Get Out of Jail Free! card, but my dad had laid down some rules, and I was meant to keep them.

"Young lady," he said, when I finally came in the front door, about an hour past curfew, "do you know what time it is?"

I pointed to all the times I did manage to keep curfew, to all the things I did right around the house and in school, but my father wouldn't bend. "That's not how it works," he said. "You missed curfew. You've got to be punished."

And I was. Graduation afternoon, all my friends headed out after the ceremony, and I had to get into the family car and go home with my parents. I'd won tons of awards at graduation, but my dad wouldn't cut me any slack. I'd broken a rule, and there had to be a consequence. To my father, it was clear. To me, it was decidedly less so, but over time I came to appreciate the absolute in this kind of justice. He set these curfews in my best interest, he always said, and here I had knowingly, willfully gone against them.

Years later, I took a somewhat different tack when I found myself coming down a little too hard on Chris for some minor transgression. Actually, it was Charles who pointed it out to me, told me I was losing sight of the bigger picture. He said, "Chris is a good kid. He's lettered in three sports, gets good grades, never gives you a whiff of trouble. You should back off on this."

And he was right. I'd lost sight of the bigger picture, so I said as much and backed off.

Indeed, there's no value in any experience, positive or negative, unless we can take something away from it. If you screw up, you screw up, but you need to learn from the mistake and move on. That was my dad's big thing, getting us kids to turn the corner on our troubles and find a better road—a place where we could move off of our missteps and onto surer footing. It's a metaphor I've used in my courtroom time and time again: that if you're really committed to turning that corner and heading down the road of hope and second chances, then after a time you'll look back over your shoulder and you won't be able to see the old road any longer.

New Hope Road. It's the road my father and mother laid out for us, the road I've tried to open up for the kids in my courtroom, and my own children as well. It's right there, on our shared road map—a little hard to find sometimes, to be sure, but if you're open to direction you'll be able to find it.

## Closing Remarks

# Call Him "Son"

Sometimes you catch a piece of wise counsel in the unlikeliest places. Or a tossed-off remark rings so fundamentally true you're left wondering how it was the same thought hadn't occurred to you a thousand times before.

I was at the West End Mall in downtown Atlanta shortly after I'd been sworn in on the bench. My appointment had been something of a big deal in and around town, because I was homegrown, a product of the community I would now serve. I don't set this out to blow smoke up my own robes but to paint the scene: My picture had been in the newspaper, I'd been interviewed on the local news, folks knew who I was, and they took turns flagging me down and stopping to chat, like we were old friends.

So there I was at the mall, minding my own business, when a kindly old man approached to press his perspective on the rookie judge. I'd been getting that a lot in those first days on the bench, and I meant to shoulder it all with great good cheer. I took it in as if it meant something, which it truly did. People would come up to me all the time, telling me they knew an aunt or a cousin, claiming we had friends in common, wishing me luck in my new position, offering their hard-earned two cents on this or that issue. Atlanta was like a small town in a lot of ways, and this was one of them, and to tell the truth I didn't mind it much at all. I actually kind of liked it, and into this gracious mix stepped this well-meaning, well-spoken, well-dressed man. He wore a sports jacket and a tie and pressed slacks, at a time when most folks dressed for the mall in jeans and sweats. He seemed to be about seventy or so, and he carried himself with kindness and dignity and purpose. He was the kind of older black gentleman I was used to seeing in church as a child—a throwback to a time when folks minded their manners and minded one another.

This older man had a thing or two on his mind, and we got to talking. He chose his words carefully, as if he'd thought them through and wanted to be sure to get his point across. What he had to say had been worrying him, that much was clear, and the opportunity to present his views to someone vested with the caretaking of Fulton County's juvenile court system must have been too good to pass up. He talked about the troubles facing children today and how the world had changed since he'd been a child.

And I listened. He talked about how it was when everyone in his community looked out for one another's children, how everyone knew his name—or if they didn't know his name they knew who he was, or they knew someone in his family, or at the very least they could place him as somehow *belonging*. What he remembered most of all was the way the adults in his neighborhood used to call him "son"—a designation of mutual respect that, now that he'd mentioned it, I realized I hadn't heard much lately. It had gone the way of black-and-white television and the black-and-white values that went with the period.

*Son, how're you doin'?*

*Son, I haven't seen your grandfather in church lately.*

*Son, there's a union job opening up down at the factory, if you're still looking for work.*

*Son, it's awfully late, you better hurry on home.*

*Son, those kids you've been running with are up to no good. Best to cut 'em loose and go about your business.*

*Son, I heard you were on your way to college.*

"Judge," he said, "that's the trouble with kids today. Ain't nobody calling them 'son' anymore."

It was a simple point, but it was so important. Forget all those statistics about kids being raised in single-parent households, about inner-city youth who grow up never knowing their fathers. It wasn't about that—or if it was, it wasn't *just* about that. And forget the cliché about arrogant inner-city youth disrespecting the old folks in their neighborhoods and the old folks in turn disrespecting the young ones, because it wasn't quite about that either. No, what this wise old man was talking about had to do with a broader concept of family than merely biological; it had to do with a connection that cut across bloodlines and reached into every church, every park, every corner store. It had to do with community, and responsibility, and a shared sense of pride and purpose. It had to do with the whole lot of us throwing in together, and rooting for our children, and pulling each other up instead of tearing one another down. It had to do with looking out for our neighbors, knowing that our neighbors in turn were looking out for us. It had to do with the redrawing of those broad family lines to where the points of connection had become no longer visible, to where a thoughtful old man felt the need to reach out to a juvenile court judge to help him do something about it.

"They might not have known my name, but they called me 'son,' " he said, for emphasis, and I found myself longing for that time when my brothers were addressed the same way, when I heard my parents or grandparents call out to a neighborhood kid as if they'd known him all along, when our comings and goings were charted by a caring community of adult role models. I wondered if I'd ever had a chance to use the phrase in my own adult experience—as a judge, as a lawyer, as a friend or neighbor or nodding acquaintance—and it pained me to realize that I couldn't think of a single such exchange. It simply wasn't a part of my experience any longer, even as it had once been a fundamental part.

Son.

It was a simple word, but like the observation behind it, it was everything. Already, in my limited courtroom experience, I'd seen enough to note that this man was on to something. Our young men weren't being called "son" anymore, not even in their own homes—and it's only gotten worse in the dozen or so years since. We've gotten to the point where we now take our young people for granted, if we even notice them at all, and it is incumbent on us to be once again about the business of addressing our children as "son" and "daughter" and reclaiming them as our own. Of taking back some of that responsibility we've irresponsibly placed on their too young shoulders. Of filling up those spaces where an absent father might have been or stepping in behind an overburdened mother or grandmother.

Think back to your own childhood. Think back to that sweet old man who used to cut your hair, or the mechanic who worked on the family car, or the fellow who rode the train to the city alongside your father. Nodding acquaintances all, and yet they all took a rooting interest in your accomplishments; they all wished you well; they all called you "son."

The tragedy of the way we now live and work and scramble is that we no longer make time for one another, and in the fallout we've lost the time for each other's children. How sad is that? How troubling? We've lost the regard for one another's children—a regard we once held so terribly dear. I grew up cloaked in that regard, and so probably did you, but now it's slipped right through our fingertips and we need to grab it back if we mean to offer a guiding hand. Without that mutual regard, we're gone, because it's not as if it just disappears; there's a disregard in its place, and you can make a strong case that that's where a lot of our troubles begin. The endless disrespecting. The shifting moral compass. The decline in societal and family values. The dreadful inability of our young people to place themselves in context, to see outside themselves, to take in any kind of bigger picture than the one smiling back at them from the laminated fake IDs they carry in their wallets.

This wise old man with his salt-and-pepper hair was completely right: For too many children, in too many communities, nobody is calling them "son." Nobody is holding them accountable. Nobody is cheering for them, or supporting them, or offering them unsolicited advice. Nobody is watching their backs. Nobody is offering them respect, or expecting it in return, and these kids desperately need that respect. Oh, yes, they do. They need it coming and going. White communities, black communities . . . it's all the same. In too many cases, in too many places, we've let our role modeling and caretaking lapse to the point where these kids are out on their own, and most of them just aren't up to it. I'm sorry, but they're just not. In our rundown inner cities, in our affluent suburbs, in our working poor rural areas, it's all the same. The shift doesn't know class or race or station. All it knows is that we've lost that shared ability to keep connected and to keep watch over our young people as if it mattered. And it does. Oh, indeed, it does.

If we expect our children to do right, we've got to do right by them, and I can't shake thinking it has to start with this right here. Let us take up this wise man's insight and call our children sons and daughters once more. Let us go back to how things were, in this one regard at least, and in so doing reach closer to how things could be again.

# APPENDIX

## Resources

A book like this can only take you so far—but happily, there are a number of agencies and advisory councils designed to help parents and children through some of the trouble spots described in these pages. They do so in confidence and with confidence, so seek them out and let them help. In most cases, I've included organizations with a national reach, but in every case you'll find someone who can at least direct you to a specific office or agency designed to help you troubleshoot your specific dilemma. There'll be dozens of similar agencies listed in your yellow pages or on the Internet, but I've had long-standing professional experience with the outfits listed below.

Please don't be shy about reaching out and asking for assistance. There's no problem too big or too small for these good people—but if we keep our troubles to ourselves, even our smallest problems can get out of hand.

Write or call—in confidence and with confidence.

### TEEN DRUG PREVENTION

Substance Abuse and Mental Health Services Adminstration (SAMHSA)
U.S. Department of Health and Human Services
Treatment Facility Locator
Resource Referral Network
www.dasis.samhsa.gov

Parenting Is Prevention SAMHSA
Informational and Resource Referral Network
www.parentingisprevention.org

### TEEN PREGNANCY PREVENTION

National Campaign to Prevent Teen Pregnancy
1776 Massachusetts Avenue, NW
Suite 200
Washington, D.C. 20036
202-478-8500
www.teenpregnancy.org

National Organization on Adolescent Pregnancy, Parenting and Prevention, Inc. (NOAPPP)
2401 Pennsylvania Avenue, NW
Suite 350
Washington, DC 20037

202-293-8370
Fax: 202-293-8805
E-mail: noappp@noappp.org
www.noappp.org

Georgia Campaign for Adolescent Prevention (G-CAPP) (Regional—Georgia)
Michele Ozumba, Executive Director
100 Auburn Avenue
Suite 200
Atlanta, GA 30303
404-475-6048
Fax: 404-523-7753
www.gcapp.org

**TRUANCY INTERVENTION**

National Dropout Prevention Center
Clemson University
209 Martin Street
Clemson, SC 29631-1555
864-656-2599
Email: ndpc@clemson.edu
www.dropoutprevention.org

Afterschool Alliance
1616 H Street, NW
Washington, DC 20006
202-296-9378
www.afterschoolalliance.org

Kids In Need of Dreams (KIND, Inc.)—Truancy Intervention Project (Regional)
395 Pryor Street
Suite 4122
Atlanta, GA 30312
404-224-4741
Fax: 404-893-0751
www.truancyproject.org/FultonTIP.html

#### TEEN VIOLENCE INTERVENTION

National Youth Gang Center (NYGC)
Institute for Intergovernmental Research
PO Box 12729
Tallahassee, FL 32317
850-385-0600
Fax: 850-386-5356
www.iir.com/nygc

Office of Juvenile Justice and Delinquency Prevention (OJJDP)
810 Seventh Street, NW
Washington, D.C. 20531
202-307-5911
www.ojjdp.ncjrs.org

#### TEEN PROSTITUTION INTERVENTION

End Child Prostitution, Child Pornography, and Trafficking of Children for Sexual Purposes (ECPAT—USA)
157 Montague Street

Brooklyn, NY 11201
718-935-9192
Fax: 718-935-9173
E-mail: info@ecpatusa.org

The Concerned Educator Allied for a Safe Environment
Project CEASE (Regional)
395 Pryor Street
Atlanta, GA 30312
404-224-4549

You Are Not Alone (YANA) Place
2013 West Pratt Street
Baltimore, MD 21223
410-566-7973

**MENTORING**

MENTOR/The National Mentoring Partnership
1600 Duke Street
Suite 300
Alexandria, VA 22314
703-224-2200
www.mentoring.org

Big Brothers Big Sisters, Inc
230 North 13th Street
Philadelphia, PA 19107
215-567-7000
www.bbbsa.org

Boys and Girls Clubs of America
1230 West Peachtree Street
Atlanta, GA 30309
404-815-5700
www.bcga.org

**CHILD ABUSE PREVENTION**

Child Abuse Prevention Network
www.child-abuse.com

Prevent Child Abuse America
200 South Michigan Avenue
17th Floor
Chicago, IL 60604-2404
312-663-3520
Fax: 312-939-8962
E-mail: mailbox@preventchildabuse.org
www.preventchildabuse.org

Childhelp USA
National Headquarters
15757 North 78th Street
Scottsdale, AZ 85260
480-922-8212
Fax: 480-922-7016

National Child Abuse Hotline
1-800-4-ACHILD

# Acknowledgments

To Dan Paisner, the wonderful talented writer, who moved this book from my heart to these pages with tremendous patience, sensitivity, thoughtfulness, and care.

To Joel Brokaw for believing in my work, in this book, and making the necessary connections to make it possible.

To Mel Berger, my literary agent at William Morris, for his tremendous effort and expertise in guiding this book to fruition.

To Mauro DiPreta, my executive editor, for his commitment and thoughtful attention to this book, and the great men and women at Harper Collins who supported this project.

To Jim Jackoway, my attorney, for his vigilant attention to all aspects of my career.

To Karen Baynes and Trenny Stovall for their uncompromising commitment to me and to the issues affecting families and children. Strong young women keep coming and you both make me very proud.

To Terry Walsh, a truly dedicated child and family advocate, whom I deeply appreciate and respect.

To Steve Mosko, president of Sony Pictures Television, for his encouragement and support, and special thanks of course to Bob Oswaks and his team for all their efforts.

To Naeemah Binion, my personal assistant, for her tremendous work in keeping all of the parts of my life in order and to Leslie Abbott-Smith, my former law clerk, for always being there in times of crisis.

And last and very importantly, thanks to my amazing extended family: Johnnetta Cole, my big sister and mentor, who has stood in the gap with me and for me in ways that have transformed my life.

Paul Hatchett and Kolen Hatchett, my brothers, and Clarence Cooper and Cecil Phillip, my surrogate big brothers, for all of your loving support in immeasurable ways.

Ethan Che Cole for being there for my sons in so many ways over and over again.

Gerald and Muriel Durley and my Providence Church family, as well as my terrific neighbors and friends who have loved and cared for my family.

The wonderful network of parents that it has been my privilege to know, especially the Tennysons and Dubins for being surrogate parents to my children, and my sons' loving godparents, Billy and Ivenue Stanley and Lydia Walker.

I am deeply grateful to my mom and sons for their unconditional love. God has blessed me beyond measure!